Conserving Biocultural Landscapes in Malaysia and Indonesia for Sustainable Development

Saiful Arif Abdullah • Amin Setyo Leksono • Sun-Kee Hong
Editors

Conserving Biocultural Landscapes in Malaysia and Indonesia for Sustainable Development

Springer

Editors
Saiful Arif Abdullah
Institute for Environment and Development (LESTARI), Universiti Kebangsaan Malaysia
Bangi, Selangor Darul Ehsan, Malaysia

Amin Setyo Leksono
Department of Biology, Faculty of Mathematics and Natural Sciences
Universitas Brawijaya
Malang, East Java, Indonesia

Sun-Kee Hong
Institution for Marine and Island Cultures and Department of Liberal Arts
Mokpo National University, Mokpo Campus
Songrim-ro, Jeollanamdo, Republic of Korea

ISBN 978-981-16-7242-2 ISBN 978-981-16-7243-9 (eBook)
https://doi.org/10.1007/978-981-16-7243-9

This Springer imprint is published by the registered company Springer Nature Singapore Pte Ltd.
The registered company address is: 152 Beach Road, #21-01/04 Gateway East, Singapore 189721, Singapore

Preface

Malaysia and Indonesia are developing countries in the Southeast Asian region. Both countries have experienced tremendous social and economic development changes, especially since the past 20 to 30 decades ago. In line with the changes, many natural areas, including forests and other natural ecosystems, were threatened and changed to other human land uses. The population of many species of flora and fauna (i.e. biodiversity) plummeted. Some were endangered and vulnerable to extinction based on the International Union for Conservation of Nature red list. Also, the loss of biodiversity affected the livelihood of local communities and indigenous people that depend on the natural resources for their livelihood.

For a long time, people from these countries have had a close relationship with biodiversity; they use it sustainably in accordance with their cultural values and wisdom or traditional ecological knowledge, which was handed down from one generation to the other. This relationship was established in many parts of the countries, which formed a unique socio-ecological system called 'biocultural landscape'. However, due to high developmental activities, the biocultural landscape is facing a serious threat and in the context of sustainable development, it is pivotal to conserve the biocultural landscape. Therefore, it is important to highlight the needs and importance of conserving the biocultural landscape and its role to the socio-economic development and the environment, which is significant to the sustainable development of the countries. In fact, such information is still lacking compared to its counterparts in European and East Asian countries.

This book volume highlights the perspectives, concepts, evaluations and assessments by researchers conducting research in Malaysia and Indonesia on issues related to the conservation, protection and preservation of biocultural landscapes. Generally, research on the biocultural landscape related to conservation can be considered still in its infancy for both countries. Therefore, the content presented in this volume's chapters is considered an important starting point in identifying issues related to concepts, theories, methods and approaches in conserving the biocultural landscape in both countries in particular and the Southeast Asian region in general. The identification aids in driving research exploration that needs to be done in the near future. However, this exploration is a challenge, given that both countries have various natural and cultural resources that contribute to the complexity in the relationship between the two resources.

Even the complexity of the relationship is displayed in the chapters of this volume, where it is essential to understand before a conservation approach is taken in a biocultural landscape. Nevertheless, the complexity is now becoming more complex because it is influenced or affected by rapid global environmental changes, especially climate change, which leads to various natural hazards, including rising of sea level, floods, droughts, storms, earthquakes and tsunamis. The natural hazards are especially impressive for Indonesia, which is a country located in the Pacific ring of fire, and this situation is presented in chapters of this volume. Additionally, the chapters in this volume have also shown that understanding the complex relationships is essential for the need to conserve the biocultural landscapes where the importance of natural and cultural resources and the relationship between them is observed to benefit local people. Such benefits are observed in social, economic and environmental contexts and clearly here encompass the three pillars of sustainable development. In this regard, the final chapter of this volume summarises the need for biocultural landscape conservation not only for sustainable development in the biocultural landscape itself but also as support in achieving sustainable development and Sustainable Development Goals (SDGs) in the countries by the year 2030.

Bangi, Malaysia — Saiful Arif Abdullah
Malang, Indonesia — Amin Setyo Leksono
Songrim-ro, Jeollanamdo, Republic of Korea — Sun-Kee Hong
August 2021

Acknowledgements

We would like to express our utmost appreciation and thanks to all the contributor authors of this book. This book is part of the collaboration between the Universiti Kebangsaan Malaysia (UKM), Bangi, Malaysia, and the Universitas Brawijaya, Malang, Indonesia, through MoU between the two universities signed in 2019. While specifically from Sun-Kee Hong, his effort in contributing to this book as an editor and author is supported by the Humanities Korea Plus (HK+) project of the National Research Foundation of Korea (2020S1A6A3A01109908).

Blurb

The biocultural landscape is an area representing a relationship that has existed for long and is interdependent between humans and nature. The relationship forms a complex socio-ecological system translated through the concept of sustainability in using natural resources through the cultural or traditional practices of the local community. Therefore, with the continuity of the concept of sustainability as its core, the biocultural landscape is a heritage that needs to be conserved for countries, such as Malaysia and Indonesia. This situation is very meaningful when the idea of sustainable development is the main agenda of the world in addressing the imbalance between the environment, economy and the society. In fact, the importance of this idea is increasingly emphasised through the 2030 Agenda for sustainable development, namely, the Sustainable Development Goals (SDGs), where heritage conservation is an aspect that needs to be highlighted.

This book is explicitly published in the context of two rapidly developing economies in the Southeast Asian region, Malaysia and Indonesia. These two countries have many similarities apart from the geographical position of being located in the equatorial region. Among these similarities are the richness of natural resources and the diversity of tribes and ethnicities that contribute to the diversity of cultural heritage and place called biocultural landscape. Generally, the content of this book includes the perspectives, concepts, analysis and assessment on the biocultural landscape sustainability issues, which are the basis for disseminating the need for biocultural landscape conservation in Malaysia and Indonesia, especially in the rapidly changing environment with issues, such as global warming, sea-level rise, drought, storms, floods and the rise of communicable diseases in which at present is the COVID-19 pandemic. Additionally, this book is a way to promote research on the biocultural landscape, especially in the context of sustainable development not only in Malaysia and Indonesia but also in other Southeast Asian countries. This book can be a reference especially for researchers engaging in heritage conservation research and for policymakers, students, conservationists, planners and protected area managers. As it is suitable to use as a reference by various stakeholders, it is hoped that this book can contribute to sustainable development efforts at the local, regional and global levels.

Contents

Editors and Contributors

About the Editors

Saiful Arif Abdullah obtained his PhD from Hiroshima University, Japan. At present, he is an Associate Professor at the Institute for Environment and Development (LESTARI), Universiti Kebangsaan Malaysia (UKM), Malaysia. His research interest includes landscape study of effects of land use changes on forested area and biocultural landscapes, forest fragmentation and their link to conservation planning and sustainability of protected area system. He has published his research works in several reputed international journals.

Amin Setyo Leksono graduated from Graduate School for International Development and Cooperation, Hiroshima University, Hiroshima, Japan. Currently, he is a Professor in the Department of Biology, Faculty of Mathematics and Natural Sciences, Brawijaya University in the field of Entomology and Ecology. He has more than 50 peer-reviewed international publications and has delivered numerous oral presentations in several international meetings. He wrote five books in Indonesian and a chapter in Landscape Ecology for Sustainable Society published by Springer in 2017.

Sun-Kee Hong is Professor at Institution for Marine and Island Cultures, Mokpo National University, Korea. He obtained his PhD in landscape ecology from Hiroshima University, Japan. His study fields are cultural landscape, island ecosystem and biocultural diversity management in Asian region. He recently published *Landscape Ecology for Sustainable Society, Biocultural Landscapes and Landscape Ecology in Asian Cultures*, all published by Springer. He is Editor-in-Chief of *Journal of Marine and Island Cultures* and a board member of International Association for Ecology. He is also a member of editorial board in Ecological Research, Acta Ecologica Sinica, Landscape Ecology and Engineering, and Biological Indicator.

Contributors

Saiful Arif Abdullah Institute for Environment and Development (LESTARI), Universiti Kebangsaan Malaysia, Bangi, Selangor Darul Ehsan, Malaysia

Elita Agustina Department of Biology Education, Faculty of Education and Teacher Training, Universitas Islam Negeri Ar-Raniry, Banda Aceh, Aceh, Indonesia

Norhayati Ahmad Department Biological Sciences and Biotechnology (Biology Programme), Faculty of Science and Technology, Universiti Kebangsaan Malaysia, Bangi, Selangor Darul Ehsan, Malaysia

Colonius Atang Faculty of Social Sciences and Humanities, Universiti Malaysia Sabah, Kota Kinabalu, Sabah, Malaysia

Jurry Foo Geography Programme, Faculty of Social Sciences and Humanities, Universiti Malaysia Sabah, Kota Kinabalu, Sabah, Malaysia
Borneo Institute for Indigenous Studies, Universiti Malaysia Sabah, Kota Kinabalu, Sabah, Malaysia

Zulfaidah Penata Gama Department of Biology, Faculty of Mathematics and Natural Sciences, Universitas Brawi-jaya, Malang, Indonesia

Luchman Hakim Faculty of Mathematics and Natural Sciences, Department of Biology, Brawijaya University, Malang, East Java, Indonesia

Sharina Abdul Halim Institute for Environment and Development (LESTARI), Universiti Kebangsaan Malaysia, Bangi, Selangor Darul Ehsan, Malaysia

Argus Faqot Faculty of Mathematics and Natural Sciences, Department of Biology, Brawijaya University, Malang, East Java, Indonesia

Nuddin Harahab Socio and Economics of Fisheries, Fisheries and Marine Science Faculty, Brawijaya University, Malang, East Java, Indonesia
The resilience of Coastal and Fisheries Villages Research Group, Malang, East Java, Indonesia

Rizha Hardyansah Master Program of Environmental and Development, Universitas Brawijaya, Malang, Indonesia

Zakaria Hashim Faculty of Architecture, Planning & Surveying, Centre of Studies for Parks & Amenity Management, Universiti Teknologi MARA, Shah Alam, Selangor Darul Ehsan, Malaysia

Sun-Kee Hong Institution for Marine and Island Cultures and Department of Liberal Arts, Mokpo National University, Mokpo Campus, Songrim-ro, Jeollanamdo, Republic of Korea

Lee Jing Institute for Environment and Development (LESTARI), Universiti Kebangsaan Malaysia, Bangi, Selangor Darul Ehsan, Malaysia

Jae-Eun Kim Institution for Marine and Island Cultures, Mokpo National University, Mokpo, Jeonallanamdo, Republic of Korea

Ibrahim Komoo Institute for Environment and Development (LESTARI), Universiti Kebangsaan Malaysia, Bangi, Selangor Darul Ehsan, Malaysia

Amin Setyo Leksono Department of Biology, Faculty of Mathematics and Natural Sciences, Universitas Brawijaya, Malang, East Java, Indonesia

Zanisah Man Anthropology and Sociology Program, Faculty of Social Sciences and Humanities, Universiti Kebangsaan Malaysia, Bangi, Selangor Darul Ehsan, Malaysia

Rita Parmawati Socio-Economics Department, Faculty of Agriculture, Universitas Brawijaya, Malang, Indonesia, Malang, Indonesia
Postgraduate Program, Universitas Brawijaya, Malang, Indonesia, Malang, Indonesia

Dina Poerwoningsih Department of Architecture, University of Merdeka Malang, Malang, East Java, Indonesia

Harsuko Riniwati Socio and Economics of Fisheries, Fisheries and Marine Science Faculty, Brawijaya University, Malang, East Java, Indonesia
Interdisciplinary Post Graduated, Brawijaya University, Malang, Indonesia
The resilience of Coastal and Fisheries Villages Research Group, Malang, East Java, Indonesia

Dg Nur Fairuzzimah Ag Samara Faculty of Social Sciences and Humanities, Universiti Malaysia Sabah, Kota Kinabalu, Sabah, Malaysia

Imam Santoso Department of Architecture, University of Merdeka Malang, Malang, East Java, Indonesia

A. Tutut Subadyo Department of Architecture, University of Merdeka Malang, Malang, East Java, Indonesia

Tanot Unjah Institute for Environment and Development (LESTARI), Universiti Kebangsaan Malaysia, Bangi, Selangor Darul Ehsan, Malaysia

Tiwi Nurjanati Utami Socio and Economics of Fisheries, Fisheries and Marine Science Faculty, Brawijaya University, Malang, East Java, Indonesia
The resilience of Coastal and Fisheries Villages Research Group, Malang, East Java, Indonesia

Erlina Laksmiani Wahjutami Department of Architecture, University of Merdeka Malang, Malang, East Java, Indonesia

Lina Asmara Wati Socio and Economics of Fisheries, Fisheries and Marine Science Faculty, Brawijaya University, Malang, East Java, Indonesia

The resilience of Coastal and Fisheries Villages Research Group, Malang, East Java, Indonesia

Wike Brawijaya University, Malang, East Java, Indonesia
Master Program of Women's Studies, Malang, East Java, Indonesia

Erna Winansih Department of Architecture, University of Merdeka Malang, Malang, East Java, Indonesia

Aniza Zairina Study Program of Forestry, Faculty of Forestry, Institut Pertanian Malang, Malang, East Java, Indonesia

Introduction: Context of the Book

1

Saiful Arif Abdullah, Amin Setyo Leksono, and Sun-Kee Hong

Abstract

This book focuses on the biocultural landscape in Malaysia and Indonesia, highly developing economies in Southeast Asia. In the context of rapid global environmental changes, particularly in this region, conserving the biocultural landscape is significant for sustainable development and attaining sustainable development goals. Therefore, as the introduction, this chapter provides the context of this book. First, the meaning of landscape and the emergence of biocultural landscape as a contemporary landscape-scale approach in heritage conservation is presented. Next, sustainability issues of biocultural landscapes due to the impact of social-economic development are discussed. Subsequently, an introduction to Malaysia and Indonesia is provided. Socioeconomic development and its significance in biodiversity and cultural diversity warrant conserving biocultural landscapes in both countries is also highlighted. And finally, the objective, purpose and content of this book are presented.

S. A. Abdullah (✉)
Institute for Environment and Development (LESTARI), Universiti Kebangsaan Malaysia, Bangi, Selangor Darul Ehsan, Malaysia
e-mail: saiful@ukm.edu.my

A. S. Leksono
Department of Biology, Faculty of Mathematics and Natural Sciences, Universitas Brawijaya, Malang, East Java, Indonesia
e-mail: amin28@ub.ac.id

S.-K. Hong
Institution for Marine and Island Cultures and Department of Liberal Arts, Mokpo National University, Mokpo Campus, Songrim-ro, Jeollanamdo, Republic of Korea

S. A. Abdullah et al. (eds.), *Conserving Biocultural Landscapes in Malaysia and Indonesia for Sustainable Development*,
https://doi.org/10.1007/978-981-16-7243-9_1

Keywords

Biocultural landscape · Heritage conservation · Indonesia · Malaysia · Socio-ecological system · Sustainable development · Sustainable development goals

1.1 Landscape, Cultural and Biocultural Landscapes

The term landscape has been debated and contested by experts (e.g. ecologists, landscape ecologists, conservationists, landscape architects, architects, planners, geographers and historians) and, on some occasions, by ordinary people. Usually, experts or researchers provide a definition or perspective of landscape based on their field or within the context of their research objective. Generally, scientists view landscape as a place of aggregated physical attributes over increasingly large areas (e.g. Forman and Godron 1986; Urban et al. 1987), such as the land and sea, in which they interact with the environment (Naveh and Lieberman 1990). However, social scientists have interpreted landscape as a medium that reflects human tradition and culture for a long time. In such a context, the landscape is a phenomenon of how people perceive and appreciate. A long-term feedback loop interaction shapes it—either positive or negative—between humans and their natural surroundings.

In recognising the human or cultural interaction with nature or landscape, the concept of cultural landscape emerged in the second half of the twentieth century. At that time, the heritage conservation started to embrace vernacular and urban and natural systems (Bridgewater and Rotherham 2019; Chapagain 2019). Occasionally, the cultural landscape is also called rural landscape (Farina 2000) and was referenced in the Operational Guideline for the World Heritage Convention (revised 1994) (Chapagain 2019). In general, the cultural landscape is the longest-term modified landscape (Farina 2000), in which the close inter-relationship between cultural and ecological values mediates its quality and function (Naveh 1995; Buckley et al. 2008). However, Chapagain (2019) argued that the concept of cultural landscape emphasises biodiversity less than the human or culture domain, although biodiversity is part of 'nature' that plays a crucial role in influencing culture or creating cultural diversity that shapes the landscape. Chapagain (2019) then mooted that a biocultural landscape approach is a reasonable means to expand the domain of cultural landscapes. Notably, before that, Bridgewater and Walton (1996) used the four principles for understanding the dynamics of the cultural landscape by Nassauer (1995) to introduce the term 'biocultural' and then articulated that most of the world's landscapes are blends of human activity with the expression of biodiversity where those types of landscapes are called biocultural landscapes. Hong (2014) stated that the term biocultural landscape was coined to represent areas that exhibit the co-existence of humans and nature because humans depend on nature through their use of biological resources (i.e. biodiversity) to exist and develop.

Generally, the biocultural landscape is extended from the concept of the cultural landscape. And the key concept of the biocultural landscape is biocultural diversity. Several definitions of biocultural diversity have been presented, such as those by

Loh and Harmon (2005), Cocks (2006), Hill et al. (2011) and Diaz et al. (2015). The definitions are similar and generally centre on the interaction among all levels of biological diversity and all cultural diversity expressions in which they can mutually adapt and co-evolve. People perceive and appreciate biodiversity and recognise cultural diversity in biodiversity conservation. Nevertheless, Bridgewater and Rotherham (2019) provided a critical perspective on defining biocultural diversity from various sources, including those by Loh and Harmon (2005), Cocks (2006), Hill et al. (2011) and Diaz et al. (2015). And they presented six key elements of biocultural diversity that can play a role in nature and heritage conservation and achieving sustainability: (1) established nexus between biodiversity and cultural diversity, (2) extensive feedback—positive and negative—between biodiversity and cultural diversity, (3) temporal and place-based elements influenced the nexus between biodiversity and cultural diversity, (4) cultural heritage may be intangible and specific to particular culture, (5) biocultural heritage is specifically related to aspects of biodiversity dependent on traditional management and (6) biocultural heritage underpins the known eco-cultural landscape (see Bridgewater and Rotherham 2019).

In this regard, biocultural landscape emerged as a contemporary landscape-scale approach for heritage conservation to consider the mutual interaction between the biodiversity and traditional/cultural heritage of the local community and indigenous people rather than merely examining them separately, as in the traditional conservation approach. This stance is related to what Hong (2014) asserted that it is necessary to strengthen and expand the evaluation of the soundness of ecosystems beyond the biological concept of biodiversity (e.g. species diversity and richness) by considering cultural diversity. A similar assertion by Chapagain (2019) stated that living things other than humans, for example, the forest and wildlife, are not necessarily limited to the 'natural world' but also closely associated with society and culture. Moreover, Agnoletti and Rotherham (2015) highlighted the importance of the research and promotion of the concept of biocultural diversity, particularly on the European continent, for the environment and human well-being, where the landscape level (i.e. biocultural landscape) is the most appropriate perspective for understanding and applying the concept. The importance of the biocultural diversity concept to conserving the environment and human well-being was also highlighted by Hong (2014). Furthermore, Mercon et al. (2019) described the key role of the concept in promoting 'safe' living and sustainability.

The biocultural diversity concept is expanding and receiving wide attention from researchers exploring it as an effective conservation tool and a means to achieve sustainability. Furthermore, there are initiatives and/or programmes, including conferences and meetings related to the importance of the connection or interrelationship between biodiversity and culture for conservation, organised by international organisations (Hong 2014; Agnoletti and Rotherham 2015; Bridgewater and Rotherham 2019). These initiatives and programmes are important because they provide a framework to further explore the concept through scientific research of science and social sciences. Environmental issues, such as climate change, biodiversity loss, land use changes, forest degradation and fragmentation, have occurred at an unprecedented rate, especially in the past four decades, and are gradually affecting the sustainability of biocultural landscapes worldwide.

1.2 Sustainability in Biocultural Landscapes

Biocultural landscapes worldwide are undergoing rapid changes in land use. The changes are becoming critical because global environmental issues, such as climate change, global warming and biodiversity loss, are another challenge to the sustainability of biocultural landscapes. Population increases, which parallel rapid economic growth, have been observed to be the underlying factors that cause the sustainability of biocultural landscapes to be under threat. In this regard, agricultural activities are intensified because agricultural technology is improving and/or new technology is introduced. Urbanisation also concomitantly occurs and gradually expands towards rural areas, which are mostly biocultural landscapes. This circumstance is significant in European countries because biocultural landscapes are predominantly characterised by the continent's landscape, especially in the countryside (e.g. Plieninger et al. 2006; Rescia et al. 2010). As global economic growth persists, a similar pattern and trends have also been observed in the other parts of the world, such as Japan, China and Korea in eastern Asia (e.g. Kim et al. 2007; Hong et al. 1995; Jiao et al. 2014; Kamada 2017).

Rapid economic growth has influenced the social, cultural and natural aspects of biocultural landscapes. Because a biocultural landscape is a socio-ecological system, any changes in its aspects affect its parts. This complex, dynamic system contains a heterogeneous land mosaic, mainly characterised by agricultural landscapes, and is associated with high biological and cultural diversity (Antorp 1997; Plieninger et al. 2006; Agnoletti 2007). The land uses that contribute to the heterogeneity include woodlands, vineyards, pastures, open fields and small farms, which are typical in European countries (Farina 2000; Oreszczyn and Lane 2000). Paddy fields, secondary vegetation, traditional settlements and grassland are typical in eastern Asia (Hong et al. 1995; Kamada and Nakagoshi 1996; Kamada 2017). Many of these traditional land uses were eliminated and/or transformed into standardised and mechanised land uses, which made the landscape more homogenous. Simultaneously, the traditional land management practices were gradually lost.

The rapid economic growth also caused rural migration. Individuals, especially the young generation, migrated to the urban and industrial areas that provided more job opportunities and attractive incentives (Rescia et al. 2010; Kamada 2017). Consequently, biocultural landscapes became depopulated, and only elderly individuals remained (Rescia et al. 2010; Kamada 2017). This situation has caused land abandonment, which decreased the socio-ecological resilience of the landscape and increased its vulnerability to disturbance, resulting in loss of biodiversity and cultural diversity values and ecosystem services (Rescia et al. 2010). In addition, the changes can degrade aesthetic values and—to some extent—cause a loss of identity (Antrop 1997).

The sustainability of biocultural landscapes is also affected by the development of the tourism industry, especially that related to nature-based tourism. Some biocultural landscapes have high aesthetic values with beautiful scenery and unique, historic landscape features promising for tourism, recreation and leisure. For example, in Europe, the landscapes of Montando in Portugal (Pinto-Correia 2000) and the Flanders region in Belgium (Antrop 1997). In Asia, the most prominent landscapes are the Philippine Cordilleras in the Philippines, listed as a World Heritage Cultural Landscape (UNESCO World Heritage Centre 2003) and the renowned 'Satoyama' landscapes of Japan (Takeuchi 2003; Fukamachi 2017). For example, many other biocultural landscapes worldwide, the Hani paddy terraces landscape in Yuanyang County in China (Jiao et al. 2014) and the Blambangan Biosphere Reserve in East Java, Indonesia (Hakim 2017), are frequently visited tourism destinations that attract national and international tourists.

Jansen-Verbeke (2009) argued that tourism development is a new challenge in conserving and sustaining the natural and traditional values and practices in biocultural landscapes. In this regard, tourism development may have a positive impact on biocultural landscapes and their residents. Essentially, it promotes the sustainable use of all the resources in the biocultural landscape in which the sustainable tourism concept is mainly applied. Additionally, tourism is a resource for improving the socioeconomics of the local community, particularly in creating job opportunities, such as being a tourist guide and local tourism operator, and other economic activities. The opportunities may prevent rural migration, especially among young people, and the risk of becoming depopulated and abandoned, as has occurred in many 'satoyama' landscapes in Japan (e.g. Kamada 2017). In addition, the encouragement and support from the government in nature-based and cultural tourism may increase the empowerment of the communities in managing and conserving their traditional and natural heritage (e.g. Hakim 2017). Such empowerment is a crucial factor to increase young people to appreciate their traditional and cultural heritage values.

Despite these promising impacts, tourism development could cause conflict between the authority and the community. An example is an income generated from the activities mainly conducted or maintained by the community only benefits the authority, for example, the local government and tourist companies (Jiao et al. 2014). The other main issue is managing the visitors/tourists at the sites because they could cause environmental and cultural degradation and biodiversity loss (Jiao et al. 2014; Hakim 2017). Therefore, management of tourists or visitors and careful planning are necessary to ensure the sustainability of the biocultural landscape's natural, cultural and traditional resources.

Because socio-ecological systems are complex, conserving biocultural landscapes is challenging. The complexity is related to the environments, economics and social—and their interaction, and its impact and influence on the cross-generational cultural and traditional values. Therefore, conserving the biocultural landscape is significant and necessary, especially for countries endowed with rich, diverse and unique biological and cultural resources, such as Malaysia and

Indonesia, especially because of the unprecedented change in the global environment.

1.3 Socioeconomic Development in Malaysia and Indonesia: Significances to Biodiversity and Cultural Diversity

Malaysia and Indonesia are developing countries in Southeast Asia (Fig. 1.1). Malaysia has two parts: Malaysian peninsular and Malaysian Borneo, with a total area of 329,847 km^2 and, as of 2019, a total population of 32.5 million people (Department of Statistic, Malaysia n.d. https://www.dosm.gov.my). While Indonesia is an archipelago, with more than seventeen thousand islands, a total area of 1.9 million km^2, and the biggest country in the region. Indonesia is one of the most populous countries in the world, and as of 2019, the total population was approximately 270 thousand million people (World Bank https://data.worldbank.org/indicator). The main islands are Java, Sumatera, Sulawesi and Borneo

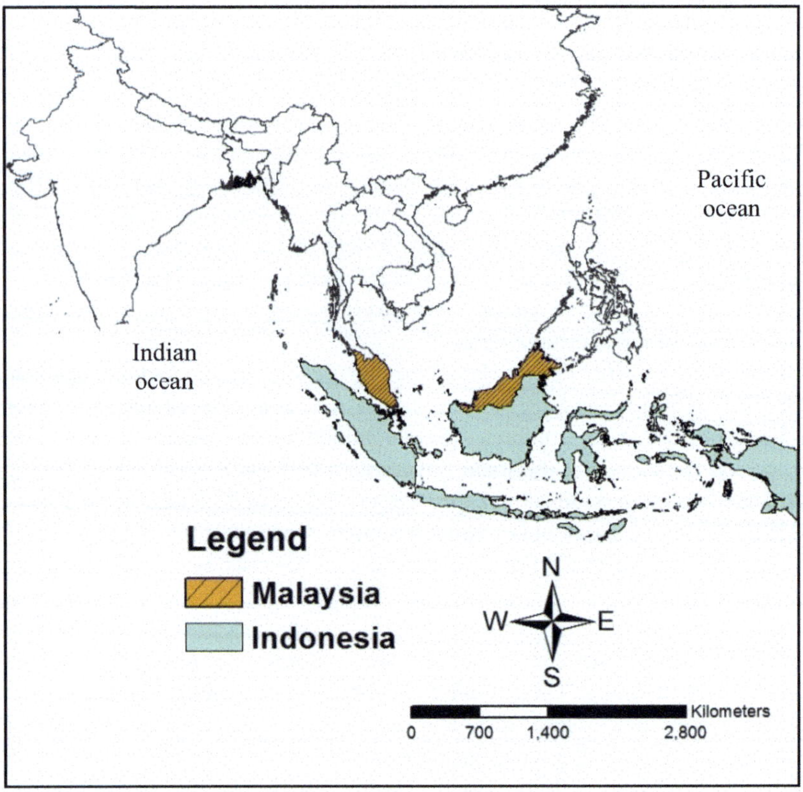

Fig. 1.1 Malaysia and Indonesia are developing economies in the Southeast Asian region endowed with biological and cultural diversity

(i.e. Kalimantan) and New Guinea. In addition, Indonesia is in a region prone to natural hazards. According to the 2017 World Risk Report, Indonesia ranked 33rd on the list of most at-risk countries to the exposure to natural hazards, such as earthquakes, storms, floods, droughts and sea-level rise. The most frequent types of natural hazards are floods (8919 instances), cyclones (5984) and landslides (4947) (Kusumasari 2019).

In the past decades, both countries have experienced substantial changes in social and economic development. Since gaining independence from British, Malaysia has improved its economy, from emphasising mainly commodity-based sectors, especially rubber and oil palm, to emphasising the manufacturing sector. This improvement propelled the country to become the leading exporter of electrical appliances, parts and components (World Bank n.d.-a https://www.worldbank.org/en/country/malaysia/overview). Its average growth rate since 2010 is approximately 5.4%, and by 2024, it could become a high-income economy (World Bank n.d.-a https://www.worldbank.org/en/country/malaysia/overview). Indonesia is the largest economy in Southeast Asia and has demonstrated impressive development since the late 1990s. For decades, agriculture played an important role in its economic growth. Nonetheless, in recent years, Indonesia's industrial and service sectors have been the main contributors to its gross domestic product. With this substantial growth, Indonesia has become an upper-middle-income country (World Bank n.d.-b https://www.worldbank.org/en/country/indonesia/overview).

Because of the mismanagement of the economic progress, many natural areas, such as dryland forests, wetlands and other natural ecosystems, are threatened and have been changed to non-natural areas. Among the world's humid tropical forests, from 1990 to 1997, those in Southeast Asia have experienced the highest rate of deforestation (Achard et al. 2004). The Food and Agricultural Organization (FAO) demonstrated that the annual rate of deforestation in insular Southeast Asia (i.e. Malaysia, Indonesia, Brunei and Singapore) was 1.75% from 1990 to 2000 (FAO (Food Agriculture Organization) 2006). In the same decade, Hansen et al. (2009) estimated the decrease in Indonesian forest cover to be 1.5% per year. From 2000 to 2010, the forest cover in Indonesia was estimated to have declined 1.0% per year, and in Malaysia, 1.4% per year, and both countries' deforestation mainly occurred in peat swamp forests (Miettinen et al. 2011).

Malaysia and Indonesia are among the 17 mega-diversity countries. These countries are endowed with very high diversity in species of flora and fauna, of which some are endemic. However, due to the loss of their natural forests and habitats, the richness and diversity of many flora and fauna species (i.e. biodiversity) have decreased significantly. Some are endangered, vulnerable and—to some extent—have become or may become extinct. For example, most medicinal plants in Indonesia identified as a priority for conservation are listed as threatened on the International Union for Conservation of Nature (IUCN) red list (Cahayanigsih et al. 2021). On the Malaysian peninsula, approximately 56% of dipterocarps taxa have been categorised as critically endangered, endangered, or vulnerable on the IUCN red list (Chua et al. 2010). Nonetheless, the loss is not limited to biodiversity. It includes environmental conditions, such as river water

Fig. 1.2 Paddy field is one of the elements of the biocultural landscape in Malaysia that provides beautiful scenery and thus aesthetic value, which are crucial for human well-being (photo by SA Abdullah)

quality and other freshwater resources being polluted by the mismanagement of development activities.

The history of human development has shown a close relationship between humans and biodiversity and other natural attributes. The nexus remains but has degraded in modern civilisation, as shown in many countries, including Malaysia and Indonesia. Today, it appears mostly in remote and/or rural areas, which are biocultural landscapes (Figs. 1.2 and 1.3). Traditionally, community members have used natural resources sustainably according to their cultural and traditional values, especially among indigenous or aboriginal tribes. The sustainability approach to the use of resources, such as for food and agriculture (e.g. river water for irrigation), has demonstrated that residents of biocultural landscapes depend much on natural resources for their subsistence and/or income generation. Thus, any degradation to the resources significantly affects their livelihood and the sustainability of the traditional management system of the resources.

About four to five decades, biocultural landscapes in Malaysia and Indonesia have been affected by rapid economic growth, and today's threat is land use development. The World Commission on Environment and Development defines sustainable development as *development that fulfils the needs of the present without compromising the ability of future generations to fulfil their needs* (WCED 1987); thus, conserving biocultural landscapes is critical in this era of rapid global environmental change. In such a context, there is an urgent need to highlight them for

Fig. 1.3 Agricultural areas of the biocultural landscape in the highland of Indonesia are prone to natural hazards (photo by AS Leksono)

conservation and their role in socioeconomics and the environment, which is significant for sustainable development and achieving Sustainable Development Goals (SDGs) by 2030.

1.4 Objective and Content of the Book

As discussed in Sect. 1.1 of this chapter, the concept of biocultural landscape is an extension of the concept of the cultural landscape. Thus, any areas identified as cultural landscapes are biocultural landscapes. Although many rural areas have the characteristics of a biocultural landscape in Malaysia and Indonesia, their identification or documentation is inferior to that of their counterparts in Europe and East Asia.

For Malaysia, only recently has such a landscape been studied. Among the earliest endeavours is Mahmud (2004), who examined the physical attributes and settings of traditional Malay settlements throughout Malaysian peninsular. Abdullah (2011) provided a general perspective of the cultural landscape characteristics within the Malaysian context. Additionally, Ahmad (2013) explored the potential cultural landscapes in Malaysian peninsular for heritage conservation and the threats and their current practices. These early studies have been crucial in understanding the biocultural landscape perspective for conserving traditional heritage in Malaysia and are an opportunity or platform to explore the human–nature relationship in a biocultural landscape context. Nonetheless, studies on the human-nature relationship have been conducted, especially in social science disciplines (e.g. anthropology and sociology), but not explicitly framed within the biocultural landscape context.

A similar situation occurred in Indonesia, but slightly advance compared to Malaysia in terms of number of studies conducted and published. Thus far, the range of topics are from conservation planning, patterns and processes to the links between humans and nature (e.g. Hakim 2011, 2017; Hong and Kim 2011; Kaswanto 2017). Therefore, to improve the gaps for conserving biocultural landscapes in Malaysia and Indonesia, this book highlights the extent of efforts undertaken to conserve biocultural landscape within a sustainable development context. To highlight the efforts is necessary because it provides a base for identifying issues of concept, theory, approach and method in conserving the biocultural landscape, particularly in the countries and Southeast Asia.

This book presents chapters on the concept of and perspectives on conserving biocultural landscapes in the context of the countries. The approaches and methods used to conserve biocultural landscapes in the two countries and the benefits to society and the environment are also discussed. The chapters also provide strategies for conserving biocultural landscapes and their implications for the related policies. There are also chapters on how the dynamic of environmental change challenges conserving the biocultural landscape in the two countries. And a concluding chapter discusses the need to conserve biocultural landscapes to achieve sustainable development and the SDGs.

References

Abdullah SA (2011) The characteristics of the cultural landscape in Malaysia: concept and perspective. In: Hong SK, Kim JE, Wu J, Nakagoshi N (eds) Landscape ecology in Asian cultures. Springer, Tokyo, pp 41–53

Achard F, Eva HD, Mayaux P, Stibig H-J, Belward A (2004) Improved estimates of net carbon emissions from land cover change in the tropics for the 1990s. Global Biogeochem Cycles 18: GB2008. https://doi.org/10.1029/2003GB00214

Agnoletti M (2007) The degradation of traditional landscape in a mountain area of Tuscany during the 19th and 20th centuries: implications for biodiversity and sustainable management. For Ecol Manag 249:5–17

Agnoletti M, Rotherham ID (2015) Landscape and biocultural diversity. Biodivers Conserv 24: 3155–3165

Ahmad R (2013) Cultural landscapes as heritage in Malaysia: potential, threats and current practice. Eburon Academic Publishers, Delft. 220 p

Antrop M (1997) The concept of traditional landscapes as a base for landscape evaluation and planning. The example of Flanders Region. Landsc Urban Plan 38:105–117

Bridgewater P, Rotherham ID (2019) A critical perspective on the concept of biocultural diversity and its emerging role in nature and heritage conservation. People Nature 1:291–304

Bridgewater P, Walton DW (1996) Towards a new framework for vertebrate conservation. In: Kitcher DJ, Suyanto A (eds) Proceedings of the 1st Eastern Ando-Australian conference on vertebrate. Western Australian Museum, Perth, pp 59–63

Buckley R, Ollenburg C, Zhong L (2008) Cultural landscape in Mongolian tourism. Ann Tour Res 35:47–61

Cahyaningsih R, Magos Brehm J, Maxted N (2021) Setting the priority medicinal plants for conservation in Indonesia. Genet Resour Crop Evol. https://doi.org/10.1007/s10722-021-01115-6

Chapagain NK (2019) Towards bio-cultural landscape: a case of Kutch. In: Chiba M, Jain S, Ghosh S, Mathur VB (eds) Cultural landscapes of Asia: understanding and managing heritage values. Dronah Publisher, Gurgaon, pp 153–159

Chua SL, Suhaida M, Hamidah M, Saw LG (2010) Malaysia plant red list: peninsular Malaysian Dipterocarpaceae. Forest Research Institute Malaysia, Kepong

Cocks ML (2006) Biocultural diversity: moving beyond the realm of 'indigenous' and 'local' people. Hum Ecol 34:185–200

Department of Statistic, Malaysia (n.d.) Online https://www.dosm.gov.my

Díaz S, Demissew S, Carabias J, Joly C, Lonsdale M, Ash N et al (2015) The IPBES conceptual framework—connecting nature and people. Curr Opin Environ Sustain 14:1–16

FAO (Food Agriculture Organization) (2006) Global Forest Assessment 2005. AO Forestry Paper 147. Food and Agriculture Organization (FAO) of the United Nations (UN), Rome

Farina A (2000) The cultural landscape as a model for the integration of ecology and economics. Bioscience 50:313–320

Forman RTT, Godron M (1986) Landscape ecology. Wiley, New York, 619p

Fukamachi K (2017) Sustainability of terraced paddy fields in traditional satoyama landscapes of Japan. J Environ Manag 202:543–549

Hakim L (2011) Cultural landscapes of the Tengger Highland, East Java. In: Hong SK, Wu J, Kim JE, Nakagoshi N (eds) Landscape ecology in Asian cultures. Springer, Tokyo, pp 69–82

Hakim L (2017) Cultural landscape preservation and ecotourism development in Blambangan Biosphere Reserve, East Java. In: Hong SK, Nakagoshi N (eds) Landscape ecology for sustainable society. Springer, Tokyo, pp 341–358

Hansen MC, Stehman SV, Potapov PV, Arunarwati B, Stolle F, Pittman K (2009) Quantifying changes in the rates of forest clearing in Indonesia from 1990 to 2005 using remotely sensed data sets. Environ Res Lett 4:034001. https://doi.org/10.1088/1748-9326/4/3/034001

Hill R, Cullen-Unsworth LC, Talbot LD, McIntyre-Tamwoy S (2011) Empowering indigenous peoples' biocultural diversity through World Heritage cultural landscapes: a case study from the Australian humid tropical forests. Int J Herit Stud 17:571–591

Hong SK (2014) Philosophy and background of biocultural landscape. In: Hong SK, Bogaert J, Min Q (eds) Biocultural landscapes. Springer, Tokyo, pp 1–7

Hong SK, Kim JE (2011) Traditional forests in villages linking humans and natural landscapes. In: Hong SK, Wu J, Kim JE, Nakagoshi N (eds) Landscape ecology in Asian cultures. Springer, Tokyo, pp 83–97

Hong SK, Nakagoshi N, Kamada M (1995) Human impacts on pine-dominated vegetation in rural landscapes in Korea and western Japan. Vegetation 116:161–172

Jansen-Verbeke M (2009) Tourismification of cultural landscapes: a discussion note. Resour Sci 31: 934–941

Jiao Y, Liang L, Okuro T, Takeuch K (2014) Ecosystem services and biodiversity of traditional agricultural landscapes: a case study of the Hani terraces in Southwest China. In: Hong SK, Bogaert J, Min Q (eds) Biocultural landscapes. Springer, Dordrecht, pp 81–88

Kamada M (2017) Satoyama landscape of Japan—past, present, and future. In: Hong SK, Nakagoshi N (eds) Landscape ecology for sustainable society. Springer, Tokyo, pp 87–109

Kamada M, Nakagoshi N (1996) Landscape structure and the disturbance regime at three rural regions in Hiroshima Prefecture, Japan. Landsc Ecol 11:15–25

Kaswanto RL (2017) Management of landscape services for Improving Community Welfare in West Java, Indonesia. In: Hong SK, Nakagoshi N (eds) Landscape ecology for sustainable society. Springer, Tokyo, pp 251–270

Kim JE, Hong SK, Nakagoshi N (2007) International trends of rural landscape researches for land management and policies. In: Hong SK, Nakagoshi N, Morimoto Y (eds) Landscape ecological applications in man-influenced areas: linking man and nature systems. Springer, Dordrecht, pp 489–504
Kusumasari B (2019) Natural hazards governance in Indonesia. Oxford Research Encyclopedia of Natural Hazard
Loh J, Harmon D (2005) A global index of biocultural diversity. Ecol Indic 5:231–241
Mahmud Z (2004) Traditional landscapes of the malays of the peninsula. Universiti Pendidikan Sultan Idris, Tanjung Malim
Merçon J, Vetter S, Tengö M, Cocks M, Balvanera P, Rosell JA, Ayala-Orozco B (2019) From local landscapes to international policy: contributions of the biocultural paradigm to global sustainability. Global Sustainability 2:1–11
Miettinen J, Shi C, Liew SC (2011) Deforestation rates in insular Southeast Asia between 2000 and 2010. Glob Change Biol 17:2261–2270
Nassauer JI (1995) Culture and changing landscape structure. Landsc Ecol 10:229–238
Naveh Z (1995) Interactions of landscapes and cultures. Landsc Urban Plan 32:43–54
Naveh Z, Lieberman AS (1990) Landscape ecology, theory and application. Springer-Verlag, New York
Oreszczyn S, Lane A (2000) The meaning of hedgerows in the English landscape: different stakeholder perspectives and the implications for future hedge management. J Environ Manag 60:101–118
Pinto-Correia T (2000) Future development in Portuguese rural areas: how to manage agricultural support for landscape conservation? Landsc Urban Plan 50:95–106
Plieninger T, Höchtl E, Spek T (2006) Traditional land-use and nature conservation in European rural landscapes. Environ Sci Pol 9:317–321
Rescia AJ, Willaarts BA, Schmitz MF, Aguilera PA (2010) Changes in land uses and management in two Nature Reserves in Spain: evaluating the social–ecological resilience of cultural landscapes. Landsc Urban Plan 98:26–35
Takeuchi K (2003) The nature of Satoyama landscapes. In: Takeuchi K, Brown RD, Washitani I, Tsunekawa A, Yokohari M (eds) Satoyama. Springer, Tokyo, pp 9–39
UNESCO World Heritage Centre (2003) Cultural landscape: the challenges of conservation. World Heritage papers 7. UNESCO World Heritage Centre, Ferrara, Italy
WCED (1987) Our common future. World commission on environment and development. Oxford University Press, Oxford
Urban DL, O'Neill RV, Shugart HH (1987) Landscape ecology. Bioscience 37:119–127
World Bank (n.d.-a) Online http://www.worldbank.org/en/country/Malaysia/overview
World Bank (n.d.-b) Online http://www.worldbank.org/en/country/Indonesia/overview

Spreading a Concept of Biocultural Diversity: From the Perspective of Island and Seascape

2

Sun-Kee Hong

Abstract

The knowledge transmitted by islanders is transformed by the continuous physical and human networking of the island. The technology and traditions of the many cultures that caught fish in the traditional way are now gone. Unfortunately, most island states are experiencing social problems due to their rapidly aging and declining populations and the effects of climate change. While the problems that the island is experiencing have to be solved through the actions of islanders, the island is only considered and viewed as relative to the land and thus has an incidental existence. The boundary between the island and the sea can be perceived as psychological, physical, and cognitive. The existence of such a boundary gives the island an imaginary and geographic possibility. It may be a desperate feeling that the islanders cannot escape forever, but it is also an escape to a new open space for land people (mainland people). Throughout the modern era, the island has become a place of development. Smaller islands were reclaimed to create larger islands, tidal flats were reclaimed to form land, and salt fields and farmland were created. In particular, some Asian and Pacific islands have been used as landfills or nuclear waste disposal sites for the mainland. Some inhabitants of the island want as much land as possible and hope that the island will be connected to the land (the mainland). Today, such a wish leads to the construction of island-mainland bridges. As urbanization progresses, islanders are demanding access to large cities. City dwellers also want to use the island as a tourist destination. Urban culture is flowing into the island through the many bridges that connect the land and the island. The biocultural tradition, which started on one island in the past, connects and propagates through various

S.-K. Hong (✉)
Institution for Marine and Island Cultures and Department of Liberal Arts, Mokpo National University, Mokpo Campus, Songrim-ro, Jeollanamdo, Republic of Korea

S. A. Abdullah et al. (eds.), *Conserving Biocultural Landscapes in Malaysia and Indonesia for Sustainable Development*,
https://doi.org/10.1007/978-981-16-7243-9_2

types of transportation and is also offset by other cultures. It is an era in which islands made up of dots (one island) connect to become a line (island to island, or archipelago) and become a face connecting with the mainland. This is not only a physical connection, but a psychological and imaginary connection that is achieved between the island and the mainland. Until now, the unique biological culture of the island has been interacting with adjacent islands and with other islands and large cities as accessibility improves. In some cases, island uniqueness disappears, and in other cases, traditions are diluted by a new urban culture. Research on biocultural diversity is no longer irrelevant to accessibility. In that sense, this chapter aims to propose the "dot-line-face model" as a multidisciplinary methodology for studying the biocultural diversity of islands and seascapes.

Keywords

Biocultural diversity · Biocultural landscape · Dot-line-face model · Identity · Indonesia · Island ecosystem · Islandness · Seascape · Sustainability · Traditional ecological knowledge

2.1 Introduction

The historical response to survival against global changes, including climate change, food problems, and resource depletion as well as the harmony and coexistence of biodiversity, cultural diversity, and sustainability, is crucial for future human survival (Hong and Nakagoshi 2017). Futurologists also regard the ocean as the Earth's last space for development. Oceans and islands are not, however, exempt from global environmental change especially given that islands are unable to cope with many environmental changes occurring in limited spaces and have inadequate resilience due to much socio-cultural vulnerabilities. Therefore, in order to solve problems related to natural resources and human life on islands and in oceans, multidisciplinary, multifunctional, and underlying research is required, research that can serve as a foundation for future policies and that goes beyond time and space and local or biased fields of research (Pretty et al. 2009; Hong et al. 2018a, b).

Human beings have already had lessons from many islands around the world. We are now reflecting on the current phenomenon of island countries suffering from the worst climate crisis and marine pollution ever. Knowledge and information from research on island and marine related climate, environment, socio-economics, natural environment, historical culture, biocultural, and policy development obtained from world island societies are required not only for island research but for the whole of humanity. A new paradigm for the sustainability of islands and biocultural diversity networks, the conservation and transmission of tangible and intangible resources, and the philosophy, thoughts, and traditional knowledge of islands and marine cultures is the future for the survival of humankind adapting to climate change. It will play a role as a "legacy of knowledge" (Clifford 1983; Hong et al.

2014). The island is a miniature version of the Earth and the future of the island is directly related to the survival of humankind.

Epistemologically and physically, islands and seas have a dual nature of openness and isolation. The sea is both an object of adventure and an object of fear. Sometimes it has been recognized as a channel for the challenge of going out into unchartered waters, sometimes as a space of ever-present and ever-changing dangers (Conkling 2007). Likewise, the island has also been perceived as a channel of communication connecting the sea, and as an isolated space separated by the sea (McCall 1994). Because of this duality, the island and the sea have been able to conceive and develop a very unique and diverse marine culture that is both communicating and changing and preserving and maintaining culture.

Looking to the future, islands and seas are becoming more important as they are perceived as alternative future spaces that will solve serious problems that humankind is encountering, such as the climate crisis and resource depletion (Hong 2020). The keywords for the future survival of humankind, the harmony of biodiversity and cultural diversity, and the solution to sustainability can also be found on islands and in the seascape. This is why human cooperation over islands and seascapes is desperately required. The seascape is an absolute space occupying two-thirds of the earth, and the land comprises a few huge islands surrounded by seas. The problems of islands and seascapes are the same as those of the earth and of human beings, the principal agents of the earth. Islands and seascapes may provide the causes of conflict and co-destruction as well as the clues to coexistence and co-prosperity. Therefore, it is necessary to combine islands and seascapes into a single category and to conduct multidisciplinary and multifunctional comparative studies that go beyond the scope of local studies and, further, to concurrently conduct underlying research that will be a foundation for policies that are directly related to the future lives of islanders.

This chapter expands the existing concept of biocultural diversity and presents a methodology related to biocultural research that can be applied to rapidly changing global environments and island ecosystems. In particular, it includes content on the application of research results. Crises such as climate change and the coronavirus pandemic are rapidly changing not only mainland but also traditional island societies' use of marine resources. From the perspective of islands and seascapes, this chapter discusses the importance of research on biocultural diversity and looks at what should be studied.

2.2 Necessity of Research on Biocultural Diversity

In order to conduct a study on the biocultural diversity of islands and seascapes, information on the diversity of bio-organisms and cultures, ecological complexity, traditional knowledge (i.e. identity) of islanders, and sustainability related to resource use is needed. Let us firstly consider "*diversity*." Looking around the islands in the Asia-Pacific, including Indonesia (Hong and Hakim 2018), I learned that there is a wide spectrum of islands in terms of nature, culture, history, and

Fig. 2.1 Seascape of Derawan Archipelago, Indonesia shows the nature, culture, history, and economy that represent the island's identity and islandness (photo by SK Hong)

economy (Fig. 2.1). As an ecologist I have particularly recognized that many academic studies have been conducted on islands. These form the basis of modern ecological theories, including evolutionary biology.

The second point is "*complexity*." Islands may be complex because they are ecologically and culturally diverse. Island ecosystems include not only terrestrial ecosystems but also various transitional areas, including seascape and tidal-flat ecosystems. Human activities are carried out not only on land but also in transitional areas. These amphibious human activities have made the island ecosystem a complex system.

The third aspect is an island's inherent "*identity*." The well-preserved terrestrial ecosystem is very stable and includes its own inherent biota. Islands also include a very unique biota if they have been isolated for a long time and their ecosystem has been preserved. This kind of island biodiversity and uniqueness depends on the island's climate, geology, and water quality. Islanders have used various biological resources on their island since they began living there. They have identified the characteristics of the soil to cultivate and engaged in fisheries according to the characteristics of the sea. As such, the characteristics of an island and its surrounding sea environment play an important accelerating role in the development of the island's cultural diversity. The agenda, "Strengthening biocultural diversity and traditional ecological knowledge in the Asia-Pacific Island region," which I worked on with the support of the Ministry of the Environment, submitted to the World Conservation Congress held in Jeju in 2012, and which was adopted as a resolution (IUCN Res. 5.115), can be seen as a result of attracting the world's attention to "islandness" as an island identity (Conkling 2007; Hong et al. 2018a, b).

Finally, we need to consider "*sustainability*." The lives and natural conservation of islanders, who depend on limited resources in a limited environment, are like two sides of the same coin. Resources are needed to live on an island but the depletion of resources has a negative impact on life on the island. Therefore, an island is a suitable place to develop a model that experiments with and implements "sustainability" (Rapport 2006). Islands are very vulnerable to climate change. Accordingly, there are many applications such as new clean energy, new industries, and new residential environmental measures (like passive houses) that are found on islands.

2.3 Methodology for the Research

Studying island biocultural diversity is an academic area where the island itself is "studied as it is" from a multifaceted perspective. The approach naturally involves the mutual intervention and collaboration of researchers from different fields, in other words, "interdisciplinary or transdisciplinary convergence." Study on biocultural diversity can be said to be an integrated academic field of sustainability to orient a consolidated approach. The origin of sustainability science can be found in the World Commission on Environment and Development report, *Our Common Future*, published in 1987. Study on biocultural diversity is both a perception system for researchers and an academic field based on the sustainability of resources and cultural continuity.

Communication between individual researchers is the basis of an interdisciplinary approach. Researchers should ideally wish to broaden, enrich, and "deepen" their specialties, bring research achievements from other fields into their specialties, and put into practice the notion that "sharing specialties is communication" as much as possible. Researchers on island humanities conduct "network-type," "participation-type," and "field-type" research activities, in addition to an interdisciplinary convergence approach. Since they cannot have in-depth knowledge about all islands in the world, they interact, collaborate, and exchange information with local researchers and institutions via networks. They also participate in various research meetings and educational programs to interact with island communities. Researchers who investigate and analyze the biocultural characteristics of islands are said to engage in "field-type" activities.

An island may be conceptualized by common features it shares with other islands (insularity): oceanicity, narrowness, and remoteness (Table 2.1). However, most islands have features of "diversity" and "double-sidedness" that cannot be defined by one or two characteristics. Accordingly, research on an island's biocultural diversity is conducted in various aspects of "the physical (or geomorphological) shape and ecosystem of islands, use of living things, history and culture, language,

Table 2.1 General features of islands

Items	Features
Surrounded by sea	Four sides are surrounded by sea
Isolation	Far away from land
Narrowness	Available area is narrow
Climate	Sensitive to climate change
Culture	Unique and indigenous culture is preserved

and economic and social structure, depending on the field of researchers. These characteristics appear as a factor of identity which expresses the diversity of the island, whether individual or complex. Therefore, in order to view and study the complexity and diversity of the island itself, a convergence studies and transdisciplinary research method is needed" (Hong et al. 2018a, b). Moreover, all participating researchers need to be open-minded to increase their efficiency.

2.3.1 Scope and Models

2.3.1.1 Scope

The island itself is isolated, but the inhabitants of the island interact with other islands and land through the sea. Therefore, island research must be recognized and executed in terms of its relationship with its surroundings. The issues of island research are wide ranging and include many angles. Issues range from the climate crisis and global economic dimensions to very small local units such as island villages, fishing villages, and resident communities. There are particular environmental problems and issues arising in certain countries or regions, such as the problem of radioactive water in Fukushima, Japan, which have begun to spread globally. With this in mind, the scope for the research is as follows:

- Global: Anthropocene, relations between continents and islands, formation of islands, classification of islands, climate change, global warming, global economic imbalance, ocean currents, marine waste, biodiversity on land and in the oceans, migration of migratory birds, migration of large fish;
- Regional: Local disputes (disputes between islands and continents, maritime territorial disputes, etc.), biological resources (fishing grounds, inflow of foreign species), comparison of ecological cultures by regional islands, comparison of islands between countries for each similar ecosystem (e.g. biodiversity, traditional ecological knowledge, tidal flat fishing, traditional fishing, agriculture, land use, dwelling patterns, shortages);
- Local: Pattern of using biological resources, ecological culture, agricultural and fishing systems (basics of life), environmental education, traditional knowledge, community; and

- Glocal (global to local, local to global): Adaptation to the era of localization, future, and information (a concept of a global community focusing on underprivileged regions and classes) (see Park 2013).

2.3.1.2 Dot-Line-Face Model

For research to spread awareness of an island's biocultural diversity, it must look for the identity of diversity and the attributes of changes appearing in the axes of time and space while maintaining the connectivity of islands (dot-line-face model) previously suggested as a methodology (Fig. 2.2). A comparative study is attempted to see how the use of sea creatures, which is essential to island life and belongs to an island's bioculture, represents distinction from an Asia-Pacific Island perspective.

Bioculture is formed by humans who have relied on natural resources. Natural resources are thus essential elements for the formation of a bioculture and they also have a complementary relationship (Maffi and Woodley 2010; Hong 2010). Regardless of whether an island is adjacent, an island whose landscape is based on tidal mudflats, an island where the water body is dominant, a large or small island, an inhabited or an uninhabited island, an island close or distant to the mainland—it is believed that features related to the origin of resources and spatial capacity affect the formation, progression, degradation, and extinction of the island's biocultural diversity (Table 2.2). In this sense, the realization of sustainability is a very meaningful concept to be used in a limited space called an "island" and suggests the direction necessary to diffuse the concept of biocultural diversity and to set research goals.

2.4 Diffusion of Traditional Knowledge to Conceptualize an Islands's Biocultural Diversity and Community Activation

2.4.1 Transmission and Creation of Traditional Ecological Knowledge

2.4.1.1 Necessity

As the environment in the island regions of the Asia-Pacific Ocean has been relatively and continuously conserved over the past decades, it has maintained its unique ecosystems. The people who adapt to their environment are rich in traditional ecological knowledge (Hong 2010, 2011). In addition, in terms of maritime history, many historical and cultural relics and stories remain. As the scope of environmental development has expanded in recent years, concerns about cultural diversity as well as biodiversity, an indicator for evaluating an ecosystem's health, are growing. This development undermines the healthiness of the landscape and the ecosystem network. Humans have used surrounding landscapes and living things as living resources for a long time and have developed new species by means of cultivation as they were needed. The use of living things has become a background to facilitate cultural diversity, such as food and residential culture, and such ecological knowledge has spread beyond the neighboring regions to the national level. Recently,

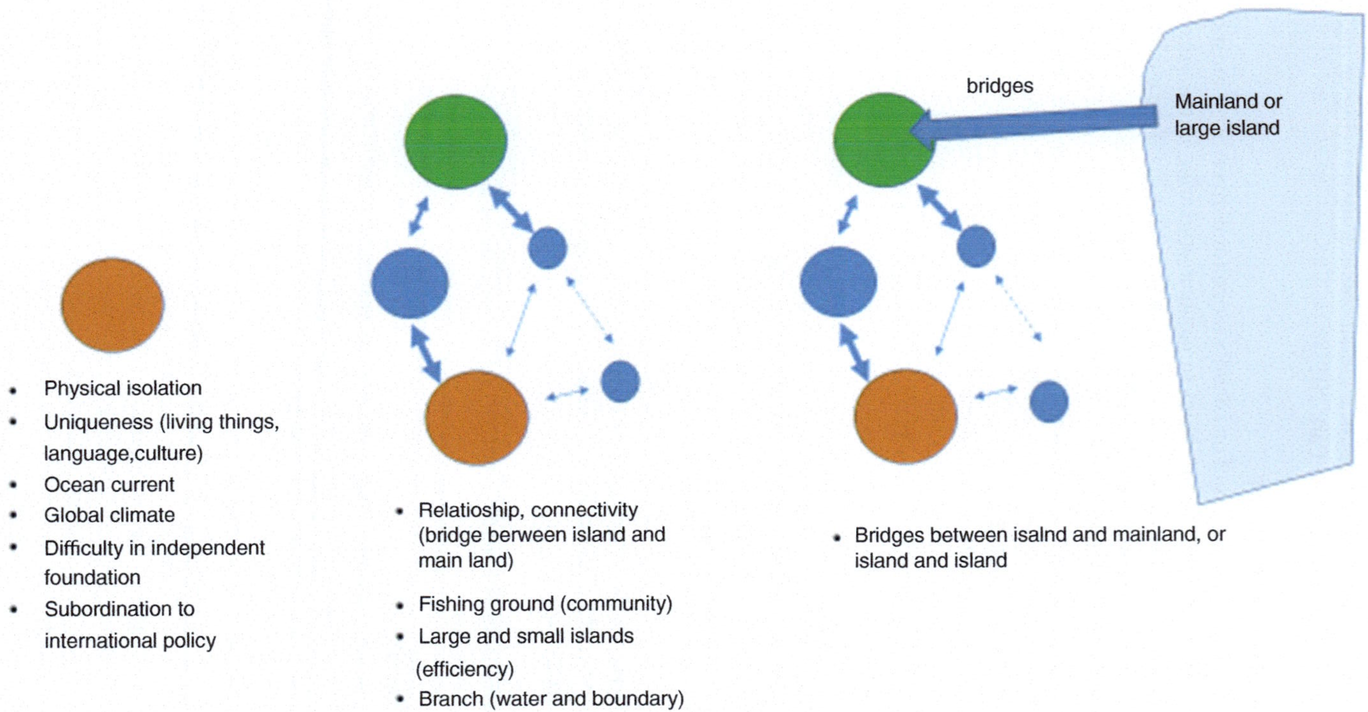

Fig. 2.2 Conceptual "dot-line-face model" diagram for understanding the biocultural diversity of island regions

Table 2.2 Research viewpoint on an island's biocultural diversity based on the "dot-line-face model" in the case of some Indonesian islands

Focus	Dot	Line	Face
Research direction	• Ecologically, environmentally, and physically isolated island (Bali, Lombok, etc.)	Island to island, archipelago (Derawan Archipelago, etc.)	Connections to adjacent cities, mainland, and large island (Madura Island, *see* my second chapter of this book)
Diversity: Biodiversity, cultural diversity, biocultural diversity, traditional ecological knowledge	• To maintain inherent biodiversity, cultural diversity, and linguistic diversity • To preserve traditional knowledge	• To share scarce biological resources, to cooperate, to diffuse social communities • To share traditional knowledge	• Influx of foreign organisms due to increase in traffic volume and tourists • Ecosystem disturbance, damage to the indigenous ecosystem of island • Cultural shock, decline in the unique cultural diversity of island • Deterioration of traditional knowledge
Complexity: Identity, insularity	• To make simple	• Increase in complexity, decline in inherent insularity and urbanization as connectivity increases due to bridge between islands • Increase in cooperative projects with adjacent islands, changes in land use (and landscape)	• Increase in cultural contact with cities, socioeconomic changes due to large bridge between island and land • Population reduction, changes in education system, possibility of increase in second and third industries (depending on location)
Sustainability: Environmental capacity, sustainable development, quality of economy, ecosystem service	• To determine sustainability of island depending on extent and resources of island	• Dispersion of artificial pressure • Increased possibility of realization of communal society (cooperation, communal society)	• Changes in island society due to outmigration (negative) • Increase in I-turn (positive) • A possibility of increased pressure on island's environmental capacity • Dispersion of pressure • Increased negative issued due to increased tourists (waste, etc.)

changes in fishing grounds that have come about due to climate change in the island regions, excessive human activities, and sudden changes in an island's ecosystem (which had been maintained for a long time and are the result of marine pollution) are going beyond affecting simply biodiversity and landscape diversity to now affect cultural diversity. In terms of the environment, this is an incident in environmental history that reminds us that the balance between humans and nature is declining and we must therefore pay it attention. Human adaptation strategies that depend on the complexity of the ecosystem are very diverse. A simple way of survival develops into unique, indigenous knowledge that depends on the methods of discovering and processing biological resources (food, clothing, and shelter) necessary for survival.

Unlike the mainland, islands, whose landscape matrix is the sea, have two-sided characteristics: isolation and openness. In many cases of ecological and cultural studies, islands and archipelagos serve as a bridgehead through which living things on continents are dispersed into the ocean as well as a filter that filters out environmental changes from the ocean to the land. Islands are an interface of biodiversity that is dispersed from the continent or originated from the ocean, and thus the related ecological information is accumulated and settles as a "knowledge system of an island's ecological culture" that forms the basis for the culture of island societies. The existence of an island ecosystem is also related to the transmission of ecological knowledge by islanders who use biodiversity. This becomes the motivating force of traditional knowledge that drives the continuity of biodiversity (Hong 2010, 2018).

Sun-dried salt, produced by using land in the tidal flat area, processed food, such as fermented fish guts and dried fish, and the use of marine biological resources, including various seaweeds, fish, and shellfish that can be obtained from tidal flats, the sea, and uninhabited islands, are the principal and underlying industries in fishing villages. An island's biological resources are used by the islanders not only as daily conventional medicine but also as major commodities to increase their income (Hong and Hakim 2018). Most plant resources on islands are medicinal plants and wild edible greens. Compared to the mountainous northeastern inland regions, plant diversity related to wild edible greens is poor, but there are native plants and trees that mainly grow and are distributed only in the evergreen broad-leaved forests on islands. Seaweed, which has been used in island areas for a long time is a resource with a high potential to be used as healthy and local food (Fig. 2.3). At the same time, however, seaweed production fluctuates largely due to climate change and environmental pollution and, as the entire island region becomes an aging society and the number of islanders capable of harvesting seaweed is reduced, seaweed is becoming a rare resource.

Traditional knowledge, which has been handed down by islanders in their use of the biological resources of islands, is very important information for studying how to preserve biological resources in the future (Chisholm Hatfield and Hong 2019). Since such traditional knowledge has complementary functions to preserve the ecosystem while using resources (Chakraborty and Alexandros 2019), it can be seen as the knowledge information system for using resources that should be preserved in terms of the national use of natural resources.

Fig. 2.3 Drying seaweed in West Sulawesi, Indonesia. Starch is extracted from the dried seaweed or used as a variety of foods (Photo by SK Hong)

2.4.1.2 Research on Traditional Ecological Knowledge

Traditional knowledge accumulated while trying to understand nature and using biological resources is an important "knowledge tradition" for maintaining an island's social community. The order of the ecosystem is changing due to recent changes in the global environment, and biodiversity is threatened by global warming. For the future survival of humankind, it is necessary to learn life strategies from past ecological knowledge and wisdom, and to pass on new ecological knowledge based on various cultural and scientific experiences.

Climate Change and Traditional Knowledge

It can be said that traditional ecological knowledge is a genetic factor that has adapted to various environments in human history (Hong 2020). Sea level rise due to climate change, the rise in sea temperature, changes in ocean currents and changes in the ecosystems of fishing grounds cause changes in living patterns of residents on islands and coastal regions (International Labor Organization 2019). The ecological factor is recognizing how the climate and environment, or the traditional knowledge inherent to islanders, has adapted to nature in the past and in the present (Chisholm Hatfield and Hong 2019). Will this cognitive system be applicable to the future island environment? To answer this, we need to explore connections with domestic and international networks for climate change and traditional knowledge.

Knowledge System on Resource Use

The long-preserved terrestrial ecosystem is very stable and has its own indigenous biota. A very unique biota also appears on islands if they are isolated for a long time and their ecosystems are preserved. This unique island biodiversity is determined by climate, geology, and water quality. Once settling on an island, islanders live on various biological resources that they find on the island. They identify the features of the soil, cultivate plants, and engage in fishing according to the characteristics of the sea. As such, the characteristics of an island and its surrounding sea environment play an important role of driving force in the development of the island's biocultural diversity.

Residents in fishing villages have fishery power, which is a unique environmental adaptation method that is not found in ordinary rural areas, and has allowed them to develop a very delicate awareness of time and space. Fishing activities depend on fishery power, whose crucial components are tide time and wind representing the tidal current. However, the lifestyle of utilizing natural resources in island areas varies as the environment of islands and oceans changes.

Biocultural Diversity and Landscape Change

Biodiversity and cultural diversity are inextricably linked. While biodiversity represents ecological healthiness and qualitative characteristics within the natural ecosystem, cultural diversity represents an important capacity for enhancing the resilience of the social system. In order to maintain the future of cultural diversity and ensure the inclusion of knowledge and creation, it is necessary to bolster the capacity of human systems to adapt to change (Kim 2013).

The use of biological resources determines people's way of life and creates culture. Land use, language systems, and the development of tools are determined by the products produced in the local area (Hong 2011). Of course, the uniqueness of these will vary depending on the size of the island and the connection between the island and the mainland, but this tradition of biocultural diversity determines the identity of the island. In recent years, the accessibility to islands has been rapidly changing all over the world due to the construction of bridges between islands and the mainland, high-speed ships, cruises, and the opening of air routes (Hong et al. 2021). Regardless of the size of the island, a lot of resources and supplies are being brought in from the outside and tourists are going to islands. These factors cause changes in the islands. It is time to carefully explore how these social and cultural shocks are affecting traditional societies on islands and how they are changing the system of biocultural diversity.

Cultural Landscape of the Sacred Place

Sometimes, the force of nature may be a *force majeure* for islanders who are dependent on the sea. To survive, they have gathered a lot of wisdom about how to read nature and live in acclimation with it. Village rituals and sacred places, including sacred hills, sacred forests, and sacred trees, play a role in prayers for peace and a bumper crop in the island villages as well as for safety on the sea. Currently, these sacred places are disappearing from the islands, and rituals are also declining.

However, "stone-wired" (a traditional fishing method that uses the ebb and flow of the tide) fishing spots are used as sacred places in some overseas island areas. Globally, the value of sacred places is acknowledged not only as important biocultural assets of humankind, but also for the role they play as media to pray for peace, safety, and a big catch. In-depth research on sacred places and their surrounding cultural landscapes and ecological features is being conducted both domestically and internationally (Fig. 2.4).

Cultural Identity of the Island's Language and Food

Nature and culture have contact points in many areas ranging from values, religion, and lifestyle to knowledge and language. The natural environment supports cultural processes, actions, religious beliefs, and the results of various cultural performances that are expressed and settled in the landscape. The most physical and chemical fusion between cultural diversity and biodiversity is in language and food. There are some key areas around the world with high biodiversity that are also very important for their cultural diversity as they have a high density of ethnic minorities and great linguistic diversity. Living things may be the same the world over, but they have different names in many cases, and their use can also be different. Food diversity is closely related to the local ecological environment. Therefore, food, recipes, and cooking methods, including food ingredients and spices, are also indicators that reveal identities of tribes and ethnic groups who have adapted to their environment.

Fig. 2.4 Sacred place and rituals in Bali, Indonesia (photo by SK Hong)

Sustainability

Culture is formed by humans who have relied on the natural resources in their environment. Thus, natural resources are essential elements for the formation of culture, and they also have a mutually complementary relationship. Even though it is an island, I believe that features related to the origin of resources and spatial capacity—an island whose landscape is based on tidal mudflats, an island where the water body is dominant, a large or small island, an inhabited or an uninhabited island, an island close or distant to the mainland—affect the formation, progression, degradation, and extinction of an island's biocultural diversity. In this sense, the realization of sustainability is a very meaningful concept to be used in a limited space called an "island."

2.4.2 Autonomy and Sustainability of Community Activities

2.4.2.1 Necessity

The global islander's use of resources, maintenance of cultural resources, preservation of the environment, and improvement of quality of life are directly related to island policy. The increased interest in island countries and island regions, which are very vulnerable to sea level rise, must include the need for international interest and assistance for the conservation of biodiversity and traditional knowledge and for the safety of islanders. Such interest is increasing in many countries around the world, but the problem of "island and biological resources indigenous to the island where humans reside, and the culture and life of residents who have been using them" is drawing less attention. In this sense, IUCN Resolution 5.115 aims to overcome this point of view, to focus on raising international interest in preserving the biological resources of islands and the culture and safety of islanders who use them, and to expand support via a global network. Building an international community network for island biological (cultural) diversity, governance, ecosystem necessary for sustainability, conservation of biodiversity, wise use of ecological knowledge, and the sustainability of cultural diversity is required. Moreover, holistic research is needed in order to carry out the core mission to develop, operate, and practice programs together with islanders.

In such a way, the vision of the "Global Island Biocultural Diversity Initiative" that I developed with the support of the Ministry of the Environment of Korea in 2016 as a follow-up strategy for the IUCN Resolution 5.115, is about "securing the sustainability of islands by preserving the universal values of nature and human heritage." This vision has a foundational strategy to maintain the maximum environmental capacity for biological and physical natural resources owned by the island itself. The fundamental background of this strategy is to ensure, as much as possible, the quality of life and cultural values of the islanders living there on the basis of biodiversity and biological resources that maintain the island landscape and surrounding waters. It also includes a foundational strategy for improving the quality of the ecosystem. In order to realize this vision, the setting of goals to be achieved both domestically and internationally must be organized in a harmonious manner.

The most important issue at home and overseas is the sustainable development of island communities. Various efforts are required to secure sustainability, but the most important is how to share and develop strategies with the residents in each country or region. To do this, education and publicity through various channels are necessary.

2.4.2.2 Research on Community Activation

Engaging and activating island communities cannot be done through the theories of researchers. Cooperation with residents is absolutely necessary. This can be achieved as follows:

Civic Education for Sustainable Development of Islands

The island is a very vulnerable ecosystem, considering its extent, resources, and isolation. Sudden changes in the lives of islanders using scarce resources are directly related to the destruction of the ecosystem. Therefore, awareness education for residents, nature conservation education, and ecosystem restoration are required in order to preserve island ecosystems.

Build-Up of a Sustainable Development Model Based on Intrinsic Development

Relying on external capital cannot promise sustainable development in the long run. It is necessary to grasp the value of the island and to think about ways to develop the region by efficiently utilizing it. Ecotourism and sixth industry, based on basic livelihoods and intrinsic development through the use of resources, taking into account the environmental capacity of the island, must be developed. An island sustainable development strategy based on qualitative economics and economic independence is needed.

Diffusion of Understanding on Harmonious Coexistence Between Human and Nature in Community Activities

Researching and disseminating the living wisdom and ecological knowledge of island communities, which use limited space and resources, allows us to respond with a counterstrategy to the resource scarcity and climate crisis that may threaten humans in the near future (Hong 2020). People around the world must keep islands not only as a territory and resource, but also as a place of peaceful coexistence, internationally sharing in the cultural, ecological, and environmental importance of islands.

The Link Between Island-Specific Future Knowledge and Sustainable Technology

Support for the development of island-specific future technology suitable for the size and environment of islands must be provided in order to build a combined circular socio-ecological system. In addition, a circular and qualitative economy that can facilitate "production on the island, and consumption on the island," to make island independence healthy and prevent environmental destruction is also needed.

2.4.3 Establishment and Application of Island Policies Based on Biocultural Diversity

The agenda, "Strengthen Biocultural Diversity and Traditional Ecological Knowledge in Asia-Pacific Island Regions" was adopted as an IUCN resolution at the 2012 IUCN World Conservation Congress (IUCN Resolution 5.115). The Korean Ministry of the Environment therefore needs to organize an active implementation program in cooperation with the IUCN to realize the agenda according to the above results. An active response with international leadership is also required for related topics in the future (Hong 2013). The IUCN Resolution 5.115, along with Small Island Developing States (SIDs), has expanded opportunities for active responses to various issues concerning the conservation of traditional ecological knowledge on island bioculture, the preservation of ecosystems, and sustainable development in mutual cooperation with island countries around the Asia-Pacific.

The IUCN Resolution 5.115 includes the concept of biocultural diversity, the use of biological resources and related traditional knowledge industries, to which the IUCN, CBD, UNESCO, and other international organizations pay attention. The increased interest in island countries and island regions, which are very vulnerable to rising sea levels, underlines the need for international assistance for the conservation of biodiversity, traditional knowledge, and the safety of islanders.

A global strategy to build an international network for global island biodiversity, governance, maintaining an ecosystem necessary for sustainability, the conservation of biodiversity, the wise use of ecological knowledge, and the sustainability of cultural diversity, and to play a key role in developing, operating and practicing programs along with the IUCN, was the "Global Island Biocultural Diversity Initiative," which has five goals as follows:

2.4.3.1 Support for Vulnerable Ecosystems

The island is a very vulnerable ecosystem, considering its extent, resources, and isolation. Sudden changes in the lives of islanders using scarce island resources are directly related to the destruction of the ecosystem. It is therefore necessary to provide awareness education for residents, education on the conservation of nature, and ecosystem restoration in order to preserve island ecosystems.

2.4.3.2 Creating Re-awareness of Ecosystem Services

Researchers must accurately investigate and describe ecosystems of islands and oceans that show high value in evaluating ecosystem services such as tidal flats and coral reefs. Coral reefs in the Pacific islands and mudflat wetlands in the regions of Asia, which are considered simply as a type of marine ecosystem, are not only ecosystems but also living places. We therefore have to publicize that they need to be evaluated appropriately when evaluating ecosystem services, specifying that they are the places where bioculture is created.

2.4.3.3 Activation of Community and Creation of Social Safety

The pace of urbanization and the decline of traditional society are closely related in the process of modernization. However, institutional mechanisms are needed for local residents to be offered economic aid because the island environment can be managed only when local residents protect the island so as to live there. Various programs that can be operated by residents autonomously, including ecotourism, cultural conservation, the introduction of the sixth industry and fair trade, may be created and guided.

2.4.3.4 Preservation of an Island's Bioculture in Response to Climate Change

The gradually disappearing island bioculture is an important legacy of humankind. As with the world's assessment on the importance of endangered species, the value of island biocultural diversity should also be evaluated as important, graded, and marked for special management (e.g. an island biocultural resource category).

2.4.3.5 Understanding the Coexistence Between Human and Nature in a Limited Space

Researching the living wisdom and ecological knowledge of islanders who are using the limited space and resources of their islands allows us to utilize them as important data for producing a counterstrategy to the various problems of resource scarcity (food, energy, water) that may threaten humans in the near future.

2.5 Conclusion

An island's identity and island identity are formed based on the physical, ecological, and cultural diversity of the island. In particular, the use of biological resources is closely related to the formation of the island's life culture. There have been various island studies conducted around the world and various methodologies for biocultural studies have been undertaken. However, the locality of an island is very important in linking the identity of the island with islandness. Therefore, since perceptions of islands and seascapes vary according to countries and regions, various research methods have been developed according to the unique regional characteristics of islands. Most of the research on the island has been conducted in the humanities, on history, culture, logistics exchange, and trade, but recently, fields closely related to the life of the islanders, especially in terms of the climate crisis, have been academically linked. Ecological, fisheries, and anthropological perspectives are making significant inroads in Island Studies. In the research so far, I have studied the viewpoints of biodiversity or biological (ecology, fisheries) and cultural diversity (ecological culture, anthropology) related to the use of biological resources for islands and coasts, which are areas that are very sensitive to global environmental changes. Through interdisciplinary methodology, the concept of biocultural diversity was explained and expanded. Although there are professional definitions of biodiversity and cultural diversity, islanders, policy makers, activists, and even

Table 2.3 Academic transboundary and interdisciplinary research fields that study nature and culture (Pretty et al. 2009)

A field of study that links the boundaries between natural and social systems	
Agricultural sustainability	Environmental sociology/ethnobiology
Biological diversity	Ethnobotany
Cognitive anthropology	Ethnoecology
Commons studies	Ethnolinguistics
Cultural anthropology	Ethnoscience
Cultural geography	Historical ecology
Cultural(landscape)ecology	Human ecology
Deep ecology	Human geography
Descriptive historical particularism	Indigenous knowledge
Development studies	Intercultural education
Ecofeminism	Landscape ecology
Ecological anthropology	Nature society theory
Ecological design	Political ecology
Ecological economics	Resilience sciences
Ecosystem health	Science and technology studies
Environmental anthropology	Social-ecological systems
Environmental education	Sustainability science
Environmental ethics	Symbolic ecology
Environmental history	System ecology

researchers express the breadth and depth of understanding of the two paradigms differently. This multi-layered and complex spectrum of recognition can make it difficult to establish a methodology for understanding and evaluating biocultural diversity. Island identity, islandness, and biocultural diversity are opposed to quantitative methods for estimating the number of species. Before the IUCN Resolution 5.115 came out, I had discussions with various stakeholders such as island-related researchers, NGOs, and university professors around the world. What I felt in this process was that although the importance of each country was recognized, it was not possible to define island culture in detail and also cover all urgent problems. For that reason, multidisciplinary, interdisciplinary, and convergent research is absolutely necessary, especially for island research.

Table 2.3 presents a study that acts as a bridge connecting academically separated systems of nature and culture (Pretty et al. 2009; Hong et al. 2014). This is an important interdisciplinary field to explain the interconnection between biodiversity and cultural diversity. It is related to the development of nature and humanities research fields related to human and social changes (cultural anthropology, environmental sociology, environmental history, etc.). Scientific knowledge (landscape ecology, biodiversity, etc.) and research fields on international politics related to the conservation of biological capital (commons research, political ecology, etc.) are central (Rapport 2006). Nature and culture intersect in many fields spanning values, religion, lifestyle, knowledge, and language (Berkes et al. 2000). The natural environment is the expression of cultural processes, actions, beliefs, and various cultural outcomes on the landscape. In this way, the connection between cultural diversity and biodiversity is physically fused through multidisciplinary research.

Acknowledgements This manuscript was written with the support of the Humanities Korea Plus (HK+) project of the National Research Foundation of Korea (2020S1A6A3A01109908).

References

Berkes F, Colding J, Folke C (2000) Rediscovery of traditional ecological knowledge as adaptive management. Ecol Appl 10(5):1251–1262

Chakraborty S, Alexandros G (2019) Community values and traditional knowledge for coastal ecosystem services management in the "satoumi" seascape of Himeshima island, Japan. Ecosyst Serv 37:1–14

Chisholm Hatfield S, Hong S-K (2019) Mermaids of South Korea: Haenyeo (women divers) traditional ecological knowledge, and climate change impacts. J Mar Island Cult 8(1):1–16. https://doi.org/10.21463/jmic.2019.08.1.01

Clifford G (1983) Local knowledge. Further essays in interpretive anthropology. Basic Books, New York

Conkling P (2007) On islanders and islandness. Geogr Rev 97(2):191–201

Hong S-K (2010) Biocultural diversity and traditional ecological knowledge in island regions of Southwestern Korea. J Ecol Field Biol 34(2):137–147

Hong S-K (2011) Eco-cultural diversity in island and coastal landscapes: conservation and development. In: Hong S-K, Wu J, Kim J-E, Nakagoshi N (eds) Landscape ecology in Asian cultures. Springer, Tokyo, pp 11–28

Hong S-K (2013) Biocultural diversity conservation for island and islanders: necessity, goal and activity. J Mar Island Cult 2(2):102–106

Hong S-K (2018) Biocultural diversity and ecological values of island and coast: suggestions for sustainable development. Minsokwon, Seoul

Hong S-K (2020) Anthropocenic climate change and traditional ecological knowledge in marine and island regions. Island Cult 55:31–56

Hong S-K, Hakim L (2018) Survey report: natural resource use, environmental management and the minorities in Indonesian islands. J Mar Island Cult 7(2):108–116

Hong S-K, Nakagoshi N (2017) Landscape ecology for sustainable society. Springer, New York

Hong S-K, Bogaert J, Min Q (2014) Biocultural landscapes: diversity, functions and values. Springer-Verlag, Dordrecht

Hong S-K, Kim J-E, Ro TH (2018a) Island biocultural diversity initiative for sustainable society in Asia-Pacific Island regions. In: Hong S-K, Nakagoshi N (eds) Landscape ecology for sustainable society. Springer, New York, pp 23–37

Hong S-K, Won Y-T, Lee G-A, Han E-S, Cho M-R, Park H-Y, Kim J-E, Samantha CH (2018b) Interdisciplinary convergence research design on island biocultural diversity—case study in Wando-gun (County) Island region, South Korea. J Mar Island Cult 7(1):12–37

Hong S-K, Hakim L, Argus (2021) Bridge and islandness: the case of Suramadu bridge in Indonesia. In: Abdullah SA, Leksono AS, Hong S-K (eds) Conserving biocultural landscape in Malaysia and Indonesia for sustainable development. Springer Nature, Singapore

International Labour Organization (2019) Indigenous peoples and climate change: emerging research on traditional knowledge and livelihoods. ILO, Geneva

Kim J-E (2013) Land use management and cultural value of ecosystem services in Southwestern Korean islands. J Mar Island Cult 2:49–55

Maffi L, Woodley E (2010) Biocultural diversity conservation—a global sourcebook. Earthscan, London

McCall G (1994) Nissology: a proposal for consideration. J Pac Soc 17(2–3):1–14
Park T-W (2013) Philosophy on the border between locality and globality and relativity and universality. Philos Cult 26:41–75
Pretty J, Adams B, Berkes F, Ferreira de Athayde S, Dudley N, Hunn E, Maffi L, Milton K, Rapport D, Robbins P, Sterling E, Stolton S, Tsing A, Vintinner E, Pilgrim S (2009) The intersections of biological diversity and cultural diversity: towards integration. Conserv Soc 7(2):100–112
Rapport DJ (2006) Sustainability science: an eco-health perspective. Sustain Sci 2(1):77–84

3 Bioregion Concept for the Landscape of Traditional Village in West Sumba, East Nusa Tenggara, Indonesia

Dina Poerwoningsih, A. Tutut Subadyo,
Erlina Laksmiani Wahjutami, Imam Santoso, and Erna Winansih

Abstract

The cultural landscape of various ethnic groups in Indonesia is very rich in variety and expression. However, these resource assets have not been properly managed and many were observed to be damaged and neglected. This means architectural and cultural landscape management in Indonesia requires integrated wisdom and understanding between short-term and long-term ideal goals. There is also a research gap in the fields of architecture and landscape, both academically and in practice. This chapter, therefore, aims to re-evaluate the architectural perspective of the traditional knowledge in Indonesia while maintaining a sustainability awareness of the planet. An example of this effort is the simultaneous identification and trace of the archipelago's cultural architecture and landscape on a regional scale. This means an ecological basis is needed as a simplification effort based on the uniqueness of the large area or region. Meanwhile, the area specificity was accurately identified based on ecology to have three divisions which are Sunda shelf, Wallacea shelf, and Sahul shelf. The Wallacea region, however, has a biodiversity hotspot with the presence of certain endemic species on each island. Therefore, this chapter also intends to emphasize the need to explore the bioregion concept of the Wallacea region in the context of preservation–conservation of Archipelago Cultural Landscape (*Lanskap Budaya Nusantara*). It also proposes the main thesis on the existence of a strong relationship between the local wisdom of Archipelago Cultural Landscape and the Wallace region through a bioregion concept. Moreover, a literature review is conducted to have a comprehensive understanding of the bioregion concept of traditional villages as

D. Poerwoningsih (✉) · A. Tutut Subadyo · E. L. Wahjutami · I. Santoso · E. Winansih
Department of Architecture, University of Merdeka Malang, Malang, East Java, Indonesia
e-mail: dina.poerwoningsih@unmer.ac.id; tutut.subadyo@unmer.ac.id; erlina.laksmiani@unmer.ac.id; imam.santoso@unmer.ac.id; erna.winansih@unmer.ac.id

S. A. Abdullah et al. (eds.), *Conserving Biocultural Landscapes in Malaysia and Indonesia for Sustainable Development*,
https://doi.org/10.1007/978-981-16-7243-9_3

cultural landscapes through the use of Ratenggaro and Wainyapu in Southwest Sumba, East Nusa Tenggara, Indonesia as case studies. The ideas in this chapter are arranged in a descriptive narrative manner based on the written sources on architecture, landscape ecology, and others on resource management in West Sumba, East Nusa Tenggara as well as the regional context of the Wallacea region. Furthermore, the megalithic architecture and traditions, dryland mixed farming traditions, and *Nyale-Pasola* tradition were selected to be discussed specifically to represent the cultural landscape entities bound by territories, places, natural environments, and cycles of time. This involved the description of these cultural landscape entities in two dimensions that correspond to the bioregion concept in the Wallacea region which are the territorial and social institutions dimensions.

Keywords

Archipelago cultural landscape · Bioregion concept · Indonesian cultural landscape · Ratenggaro Village · Wainyapu Village · Wallacea region · Wallacea bioregion · West Sumba

3.1 Systemic Problems of Traditional Villages in Sumba, East Nusa Tenggara

The Sumba traditional village in the East Nusa Tenggara represents the cultural and architectural landscape phenomenon in the Wallacea region due to its rich megalithic heritage and valuable biodiversity. A clear concept is, therefore, needed to map this potential to promote the vision of preservation and conservation activities. The village has only been sufficiently discussed in the physical such as land use, spatial patterns, green patterns, etc. as well as the visual aspects such as the building architecture, building arrangements, and village patterns. Moreover, the Southwest Sumba Regency is an example of the area included in the category of Disadvantaged Areas based on Presidential Regulation Number 131/2015 concerning the Determination of Disadvantaged Areas for 2015–2019. Meanwhile, the Rattenggaro and Wainyapu villages globally known for their natural beauty and culture are located in this regency, and data retrieved in 2016 showed the potential for tourist visits to the area is quite high at 55.5% (Department of Culture and Tourism of Southwest Sumba Regency). However, the cultural landscape of Sumba provides few economic benefits to its people unlike Bali (Vel 2010; Hadi 2018).

The phenomenon of low public welfare in the East Nusa generally indicates a crisis in people's mindsets and behaviors. This is based on the lower availability of extensive natural resources and socio-economic growth compared to the other regions in Indonesia (Critical Ecosystem Partnership Fund 2014). Moreover, the economic growth in the Wallacea region averaged 7.2% per year in 2010–2012 and this was consistently higher than the national average of 6.2% at the same period (Critical Ecosystem Partnership Fund 2014). This means the region has a large contribution from the natural resources but the situation is different in East Nusa Tenggara which had only 5.46% in 2012 compared to the national average of 6.03%. In 2020, it was only 3.06% compared to the national average of 5.04% and this is evident in how people manage their natural resources, respond to food insecurity situations, deal with natural disasters, and implement weak mitigation efforts. Furthermore, poverty has also been correlated with environmental problems in this region due to its ability to cause forest degradation and the subsequent decrease in biodiversity (Vel 2010; Bashari et al. 2014, Hadi 2018; Takandjandji 2015).

The high poverty rate is influenced by the incompatibility of development programs with the economic procedures and behavior of the Sumba community (*Uma*) which Vel (2010) calls the *Uma* economy. The *Uma* economy is the production, distribution, and consumption activities of the *Sumba* local economy driven by or integrated with exchange relations between the consanguine of *Uma*. It also indicates the process of joint efforts of Sumba society to select and reproduce the mode of food exchange without having to beg (*mandara uang)* for food sovereignty. It is, however, difficult to increase the productivity of agricultural products without understanding the existence of the *Uma* economic culture. According to Laksono et al. (2019), this economic behavior is a cultural phenomenon of kinship which serves as the background for poverty in West Sumba.

The food security crisis in East Nusa Tenggara, especially the island of Sumba, is an ironic phenomenon due to the fact that the people inherited their knowledge of food security based on a socio-ecological perspective. The economic problems, especially in rural areas, are exacerbated by the complexity of rural socio-ecological affairs. Therefore, the mid and long-term solution proposed was to re-strengthen the capacity of local villages based on local knowledge related to food security. The programs need to be able to integrate the purpose of sustainability into local community values including those associated with the *Uma* economic system. It is, however, important to note that the *Uma* has socio-ecological aspects inherent in the place (*genius loci*) and local wisdom of the indigenous village community (Njurumana and Prasetyo 2010; Amin et al. 2016; Hadi 2018).

3.2 Towards the Archipelago Cultural Landscape

This chapter explores the bioregion concept of the Wallacea region in the context of preserving the conservation of the Archipelago Cultural Landscape (Santoso and Widiningtyas 2019). The ideas in this chapter are arranged in a descriptive narrative manner based on the written sources on architecture, landscape ecology, and others on resource management in West Sumba, East Nusa Tenggara as well as the regional context of the Wallacea region as indicated in Fig. 3.1. It also proposes the main thesis on the existence of a strong relationship between the local wisdom of Archipelago Cultural Landscape (*Landscape Budaya Nusantara*) and the Wallace region through a bioregion concept. A descriptive analysis was conducted by explaining the relationship between the existence of tangible heritage such as the architecture and ancient sites and the intangible ones including the traditions,

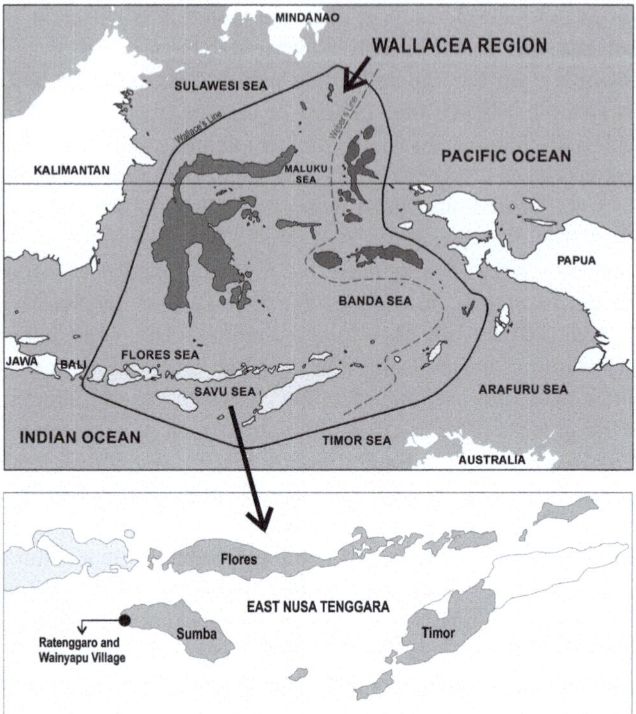

Fig. 3.1 Map of the Wallacea Region and case study location

customs, and beliefs in line with the Wallacea biodiversity existing today in Sumba. The analysis was narrative with the focus on the spatial function and meaning which were specifically selected because of their proximity to architectural and spatial substances required for landscape planning and designing. The ideas in this chapter adopt a framework developed by Zasada et al. (2017) which integrates two main dimensions including the (1) socio-institutional and (2) territorial for the development of cultural landscapes. This is believed to have the ability to ameliorate the general conceptual understanding of how policy actions impact the environment and landscape as well as human well-being. The idea of these dimensions is consistent with the understanding of the importance of a comprehensive bioregion concept for traditional villages as a cultural landscape (Poerwoningsih et al. 2018).

The United Nations Educational, Scientific, and Cultural Organization (UNESCO) appraises Indonesia to be a superpower country in the cultural field. The Permanent Delegation of the Republic of Indonesia to UNESCO (2017) showed nearly 600 Intangible Cultural Heritages have been recorded from the country with some recognized by the international body. These cultural products, however, have the potential to strengthen and enrich the character of the nation towards ensuring a better level of sustainability. It is also important to note that the number of Tangible Cultural Heritages recognized is very few compared to the potential resources. For example, only 8 sites of the country's cultural landscape which are tangible cultural heritages have received recognition as World Heritage. There is, therefore, the need for strategic efforts to manage these sites to obtain benefits for the nation's welfare. This management includes a system of categorization, recording, inheritance, and preservation for sustainable purposes. Moreover, the promotion of the cultural landscape recognition in the archipelago to the world level is not only for the preservation of the region but also to achieve more sustainable goals. This recognition is also correlated with an attitude of pride, a sense of belonging, and responsibility for the settlement area. It is also important to note that attitude is the essence of true sustainability (Wahl 2018).

The cultural landscape is a natural environment designed by human culture and a more operational definition by the UNESCO (Operational Guidelines for the Implementation of the World Heritage Convention 2012) showed it is a representation of the collective work of humans and nature. This means it has three main elements which are the natural environment, culture, and human beings placed in the middle. The cultural landscape is a concrete and distinctive product of an interaction between human communities with certain cultural preferences and potentials as well as a series of natural conditions. It possesses a long-time dimension inherited from several generations of human behavior and periods of natural evolution and also serves as a picture of the evolution of society and its settlements for adaptation to different influences and constraints caused by environmental conditions. Meanwhile, the local wisdom asset in the cultural landscape is the target to be recognized as a world heritage (Taylor 2009; Taylor and Lennon, 2011; Ardiyansyah 2014; Amin et al. 2016).

The threats to the cultural landscape need to be mitigated through multidimensional preservation and conservation activities. Meanwhile, the thinking of this chapter is strongly influenced by 2 (two) important points with the first being the principle of sustainability demonstrated by the synergy relationship between the cultural landscape and the tourism sector. It is possible to conserve the natural and artificial (cultural and historical) landscapes in line with the direction of the eco-friendly tourism being developed. This means cultural landscapes hold strategic values in driving the potential for tourism landscapes with economic and social value due to their ability to synergize with history, activities of daily life, and livelihoods inherent in natural and artificial settings. Therefore, tourism promotion content needs to be able to fully inform the synergistic spirit, while cultural landscapes serve as the source of inspiration in planning and designing landscape architecture (Taylor and Lennon 2011; Handini et al. 2016; Hadi 2018; Handini 2019).

The second point is the conservation of the archipelago cultural landscape which requires support from the protection and spatial planning aspects. The policies established for the protection of cultural heritage provide assurance which allows the formation of tiered networks ranging from the local, regional, national, and even to the international scope. This network is further used to support decision-making and investment in cultural heritage activities. Moreover, Regional Spatial Planning plays a very essential role in Indonesia and several laws have been implemented to regulate the protection and preservation with the cultural heritage regulated in Law

No. 11/2010 concerning cultural heritage, while natural heritage is regulated through Presidential Decree No. 32/1990 concerning protected areas (nature) and Law No. 26/2007 on spatial planning. In principle, the Regional Spatial Planning law regulates the structure and pattern of space with the structure focusing more on the arrangement of residential centers, infrastructures, and facilities network systems, while the spatial pattern is more on the function of protection or cultivation. Therefore, the conservation areas, both natural and cultural landscapes, need to be protected due to their specific and distinctive spatial patterns (Fredholm 2017).

3.3 Bioregion Concept for the Archipelago Cultural Landscape (Lanskap Budaya Nusantara)

The challenge was to focus more on the context of the local environment of the archipelago in line with the cultural landscape. The local context is directly related to the myths, legends, and folklore and also to appropriately participate in rituals and practices in order to maintain local diversity and resilience. Moreover, the realization of recognition for the Archipelago cultural landscape is not just to emphasize the interaction between humans and nature nor is it merely the concept of a balanced relationship between the local community and the environment. It is also needed as a reference to a measurable scope in more concrete dimensions and this led to the need for the concept of a bioregion (Hidayatun et al. 2014; Amin et al. 2016).

The realization of a balanced relationship between the local community and the environment needs to be conducted by the community in accordance with the spirit of bioregionalism. Meanwhile, the bioregion concept in this paper was inspired by the spirit of bioregionism developed by Raymond Dasmann in the 1970s. This concept further strengthens the theoretical foundation and potential to build greater cohesion between environment and culture. It is a social movement which seeks to restore a sense of place and community and also a movement to revitalize an ecologically and culturally sustainable society in the context of its local geographic area. This concept was developed based on the interests in the direction of balance and this means it provides a direction for community social change to protect and restore the environment at the local level. Moreover, bioregionalism encourages the formation of networks between community groups to make social and ecological changes at the same time (Ryan 2012; Amin et al. 2016; Evanoff 2017; Wahl 2018).

Bioregion shows a geographic space containing a single ecosystem (Miller 1996) which, in simple terms, means natural space (Ryan 2012). Meanwhile, Wahl (2018) concluded three important things to understand the concept of the bioregion and the first is to place it in the context of a planet (global) and emphasize its uniqueness and intrinsic value. The second shows the bioregion boundary is mainly determined by the local quality of the natural self-regulating process and this is related to the presence of certain local cultures and types of settlements. The third is the identification of bioregions as geopolitical entities with the residents required to be sensitive to several local environmental conditions. Moreover, bioregion establishes the responsibility of identifying the territory to be lived. Therefore, in-depth knowledge of natural cycles, flora and fauna, and sensitivity to the disturbance of the ecosystem are essential for human survival. This means bioregion has a function of human involvement through cultural activities.

The diversity of the cultural landscape in the archipelago requires a flexible and adaptive development framework with the unique local characteristics of the particular region. Therefore, the bioregion concept aims to find a balance between people's needs for livelihoods and the potential for natural resources in their area. This balance is, however, based on ecological, economic, and social criteria and this means spatial policymakers, architects, and landscape planners need to be sensitive to the bioregion planning related to the development of the cultural landscape in the archipelago. It is important to note that the planning based on the bioregion concept is an organizational process which allows stakeholders to work together to obtain information, identify problems and their potential areas, determine goals and objectives, determine the form of activities, conduct activities, take action with the community, evaluate progress, and improve activities. The bioregion concept also provides a substantive meaning for the geographical space including several ecological features such as water sources, food production land, forests as well as other natural environments and their infrastructure. It also has economic and social significance in the form of permaculture activities, community-based agriculture, cooperation between workers and consumers, community funding, local currency, ecovillages, and others (Miller 1996; Vel 2010; Lockyer and Veveto 2015).

The sustainability of the cultural landscape in the archipelago needs a framework for future research and policy design. This chapter adopts the framework of Zasada et al. (2017) and integrates two main dimensions which include (1) socio-institutional and (2) territorial to develop cultural landscapes. The territorial context relates to the effects of spatial policies which have targets and scale required to be in line with effective policy objectives and environmental impacts. It also relates to the benefits of the landscape services according to the accommodative and flexible biophysical capacity. The integration of these two dimensions is believed to be able to sharpen the general conceptual understanding of how policy actions impact the environment and landscape as well as human well-being. This point is in line with the understanding of the importance of a comprehensive bioregion concept for traditional villages as cultural landscapes as shown in Fig. 3.2.

The territorial dimension plays an important role in the mechanism to implement policies and create benefits from the landscape. Meanwhile, the local biophysical conditions including specific geographical, climatic, hydrological, and soil properties generally determine the land use as well as the productivity of food land in particular. Moreover, the property of territory also determines the potential for wider regional development concerning the cultural landscape as well as the specific land-use systems and management boundaries, especially in ecologically sensitive areas. Most of the ecological problems in customary villages are related to sensitive locations such as highlands, riparian areas, forest areas, as well as conservation areas, both in urban, rural, and rural-urban fringes. The socio-economic context refers to the demographic and economic situation of the local commodities in the region, while the biophysical context affects landscape management and

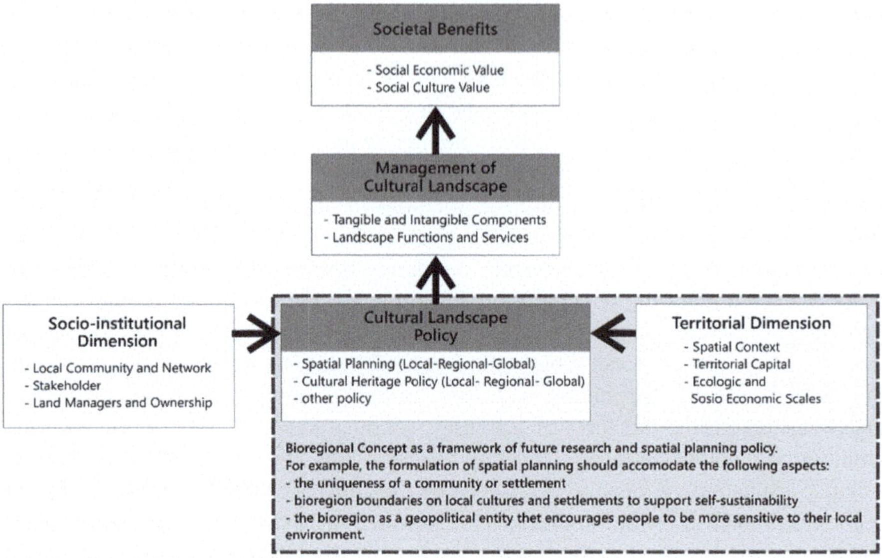

Fig. 3.2 Integrating bioregion concept in territorial dimension as a framework of cultural landscape policy

environmental policy implementation (van Berkel and Verburg 2011; Poerwoningsih et al. 2015; Zasada et al. 2017).

3.4 Wallacea Bioregion as a Regional Framework for the Archipelago Cultural Landscape (*Lanskap Budaya Nusantara*)

Wallacea is a region with high biodiversity characterized by several islands exclusively inhabited by diverse endemic flora and fauna. The riches of the region are due to geological processes caused by the movement and friction of the continental plates. The distinctiveness of the area between the Wallacea and Lydekker lines is indicated by the characteristics of the landscape system components, water infiltration, the process of land formation, diversity of flora and fauna species as well as the cultural wealth of humans that have inhabited the Wallacea region from generation to generation. It is located in the middle of Indonesian Islands including Sulawesi, Maluku, Sumba, Sumbawa, Lombok, and Timor Islands. Meanwhile, the idea of integrating the Wallacea region into the future of cultural landscape in Sumba traditional village is believed to be able to lead to two-way development on a territorial scale, both globally and locally (Bisjoe 2015; Aziz et al. 2018; Santoso and Widiningtyas 2019; Rheindt et al. 2020).

This chapter, therefore, proposes Wallacea as a regional framework for the archipelago cultural landscape in the region in line with the three important points addressed by Wahl (2018) in understanding the concept of a bioregion. The Wallacea bioregion has the potential to become a strategic framework in the management of the Archipelago Cultural Landscape towards achieving World Heritage recognition. This is in line with the findings of the Critical Ecosystem Partnership Fund (CEPF) that Wallacea is one of the world's centers of biodiversity with 391 important Key Biodiversity Areas or KBAs. These include 251 on land, 74 in water, and 66 on well-preserved marine ecosystems scattered in the Banda Sea, Halmahera, Nusa Tenggara area, the Makassar Strait, and Tomini Bay. The CEPF also found 560 species threatened with extinction based on the IUCN classification in the area. Moreover, the Sumba Island was found to be an endemic bird area with 13 bird species discovered to have limited distribution and seven of them are "puyuh sumba" (*Turnix everetti),* "punai sumba" (*Treron teysmanii),* "walik rawa manu" (*Ptilinopus Dohertyi),* "wangi/hantu sumba" (*Ninox rudolfi),* "julang/rangkong sumba" (*Rhyticeros everetti),* "sikatan sumba" (*Ficedula harteti),* and "burung madu sumba" (*Nectarinia buettkoferi),* while the sub-endemic species include "kakatua kecil jambul kuning" (*Cacatua sulphurea citrinocristata)* and "burung bayan" (*Eclectus roratus cornelia)* (Bashari et al. 2014; Bisjoe 2015; Takandjandji 2015; Santoso and Widiningtyas 2019; Rheindt et al. 2020).

The Wallacea region has three bioregions which retain their cultural values and these include Sulawesi, Maluku, and Nusa Tenggara. These values are partly related to ensuring the protection of biodiversity and its habitat in a sustainable manner. According to the Culture Statistics of the Ministry of Education and Culture (2016), the Wallacea region has 1334 cultural heritages with the East Nusa Tenggara as part of the Lesser Sunda Islands contributing 2 cultural heritages, 11 museums, and 101 intangible cultural heritages. Meanwhile, the traditional villages in East Nusa Tenggara are under threat of destruction and extinction as is the case with biodiversity due to the fire hazard which quickly destroyed dozens of traditional houses, and the renovation and reconstruction process was observed to be expensive. The road map to submit the customary villages in East Nusa Tenggara to be considered a cultural heritage landscape is, however, expected to overcome the problems and limitations of resources and funding (Takandjandji 2015; Santoso and Widiningtyas 2019).

The traditional villages scattered on the island of Sumba are administratively located in the Province of East Nusa Tenggara which is geographically an archipelago consisting of small islands. Morphologically, the Sumba Island is mostly hilly and mountainous with an average slope of 50%, while the natural character of East Nusa Tenggara includes savanna forest and a small part of tropical rainforest. It is important to note that the limited conditions of these savanna forests influence the ability of the people to manage and utilize the existing biodiversity in their way. Moreover, the agricultural model generally developed is mixed dryland farming which involves combining the potential of agriculture, livestock, forestry, and the environment. The *mamar* system was developed by the people of the island of Timor, Alor, and Rote, while the *amarasi* system is on Timor Amarasi Island, the *kaliwu* system on Sumba Island, and the *sikka* system on Flores Island. Furthermore, Sumba Island is geographically influenced by a semi-arid climate (monsoon) and this leads to the formation of a grassland environment (*sabana, savanna)* with a little thicket forest. It was also discovered that almost the entire surface of the island is a limestone layer and this means it does not allow a large variety of types of plants to grow. The term West Sumba in this paper represents a geographical region characterized by the physical boundaries of the area in the west-south location of the island of Sumba. Some geophysical differences were also observed between the north-east and the west-south parts with the north-east dominated by wide lowlands, gentle hills, and low rainfall, while the west-south has mountains with sharp ridges, lots of forests, and higher rainfall. However, there are agricultural lands traditionally managed outside the residential area in the West Sumba region even though it has arid natural conditions (Njurumana and Prasetyo 2010; Takandjandji 2015; Amin et al. 2016; Handini et al. 2016).

3.5 Megalithic Architecture and Culture in the Traditional Village of West Sumba

The architecture in the traditional village of West Sumba is in the form of culture and tradition passed down from generation to generation. It is a reflection of life involving several cultural elements such as those associated with ideology, customs, kinship systems, technology, livelihoods, and ecological resources. This means the culture is reflected in the architectural form of buildings, the structure of buildings, elements, and residential spaces which form certain land-use patterns in a particular area. Several studies on architecture have shown information on its relationship with culture, natural resources, and climate. Moreover, the *Marapu* belief underlying the way of seeing the world (cosmology) also influences the form, space, and meaning of architecture. This shows the megalithic tradition is synonymous with the presence of gravestones and the placement and stone character of these components further influence the spatial pattern of the West Sumba traditional village. They also form the fabric of resources and determine the shape of the cultural landscape as indicated in Fig. 3.3 (Mross 2000; Adams 2007; Hariyanto et al. 2012; Geria 2014).

This means *Marapu* belief is very influential in all aspects of life for the people of Sumba including the village arrangement and residential architecture. They still adhere to this belief system up to the present time even though some residents have embraced other religions such as Christianity and Islam. It is important to note that it is a combination of animism, spiritism, and dynamism which places the spirit of the ancestors (*marapu*) as the main component. Moreover, *Marapu*'s conception of belief includes the system of life in the universe which involves an eternal harmony of balance between *Ina Kalada* (great mother) and *Ama Kalada* (great father). This binary balance pattern is manifested in the system of life, arrangement of buildings, and arrangement of settlements. The people of Sumba believe in eternal life through death such that the dead become spirits which remain and become part of life in the village. Therefore, it is possible to see the burial area placed in the village environment (*parona).* The spirit is believed to reside on the roof of the towering pyramid which is not allowed to be inhabited by living people and usually

Fig. 3.3 Megalithic villages of Ratenggaro and Wainyapu (2020 documentation)

used to store ancestral heirlooms as indicated in Fig. 3.4 (Mross 2000; Kusumawati et al. 2007; Djawa 2014; Messakh 2014; Solihin et al. 2018; Agustin 2019).

The people of Sumba have a unique culture of living in three types of houses with different functions and these include the traditional houses (*uma adat),* hamlet houses (*uma dusun),* and garden houses (*uma kebun).* The traditional house (*uma adat*) functions as a place to conduct religious ritual activities and is usually located in the *parona* (village/traditional village). The hamlet house serves as a place to live every day, while the garden house is a place to live when working on the garden or farm. It is important to note that the architectural form of these houses is very different as indicated by the character of the tower (*mbatanggu)* on the roof. The traditional tower is 15 m high and only found in traditional *uma* buildings or *Uma adat* (*Uma kelada)* which is also often referred to as *uma bokulu* (big house) or *uma*

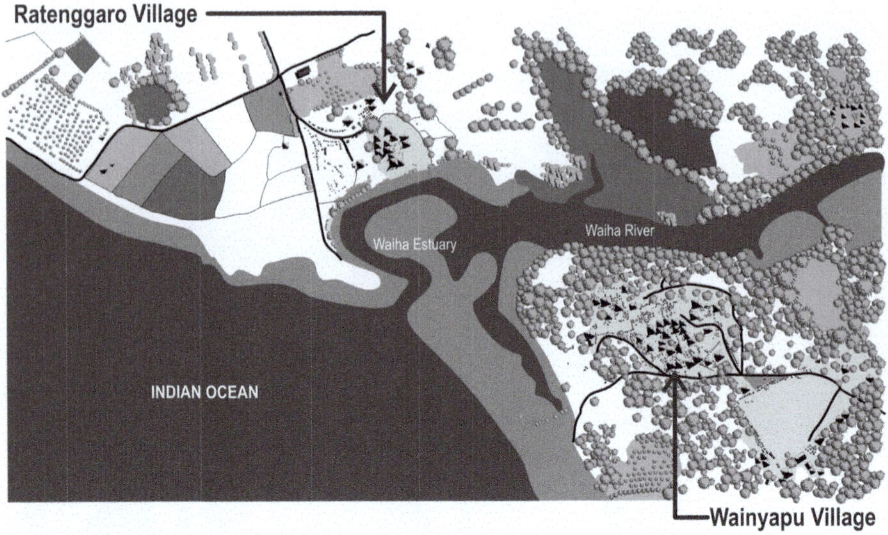

Fig. 3.4 Ratenggaro and Wainyapu villages (sketch-2020)

mbatanggu (tower house) (Kusumawati et al. 2007; Hariyanto et al. 2012; Reny et al. 2018).

The uniqueness of the culture of this settlement reflects in the territorial dimension used in regulating the different functions of these types of houses. Their locations also differ based on their placement towards the main village (parona) such that it is possible for a hamlet house to be inside a parona called *ana uma* (when built in the same village) but does not have a function like *uma ouma* (built outside the traditional village). *Ana uma*, however, means "child of the house" which is a house founded by younger ancestors. Meanwhile, garden houses are often referred to as *uma ouma* which are outside the parona or built around rice fields and fields for surveillance purposes. For the *Kodi* ethnic group in West Sumba, each family has at least two houses (*uma*) including the *uma* adat in *parona* and either of *uma dusun* or *uma kebun* (Kusumawati et al. 2007; Hariyanto et al. 2012; Reny et al. 2018).

3.6 Traditional Agroforestry and the Lende Ura Philosophy of the Southwest Sumba People

This section is based on the thesis associated with the existence of a megalithic society characterized by an agrarian lifestyle. The location of megalithic civilization sites in different places has a strong attachment to natural resources which are basically attached to agrarian tradition. Archeological findings in the form of mortar, metal tools, and others are related to farming activities, while the *Poaceae* pollen findings are from the excavated soil samples at megalithic sites in Sulawesi. *Poaceae* is a species of flowering plants such as rice, wheat, and maize which are important agricultural commodities. It is also important to note that agricultural human culture is very dependent on the availability of foodstuffs produced by these species (Hasanuddin 2015).

The territorial dimension and its properties were also currently found in the agrarian life of the people, especially in the West Sumba region. This is evident in the presence of agricultural lands traditionally managed outside the residential area despite the arid natural conditions of the area. The landscape of the Sumba traditional village also displays significant spatial heterogeneity in its biophysical and socio-economic context. Moreover, the communities recognize the traditional agroforestry system called the *Kaliwu* or *Kalego* system with the prevalent types of plants commonly grown found to be perennials such as coconut, coffee, jackfruit, candle-nut, bamboo, mango, *rambutan*, soursop, and others. The community is also familiar with field farming (*oma dana)* usually conducted during the rainy season to grow crops such as upland rice, sweet potatoes, corn, chilies, papaya, taro, rote corn/sorghum, large pumpkin, and long beans. In a small part of the area below the hill or part of the valley, the traditional people also recognize rice field farming (*paba)* which is cultivated during the rainy season. These forms of agriculture and agroforestry, therefore, characterize the land use in the West Sumba region and the landscape of the traditional village of West Sumba (Njurumana and Prasetyo 2010; Amin et al. 2016).

The socio-institutional dimension can also be reflected through the participation of the local community in nature conservation efforts. The traditional communities around the *Gunung Yawila* Protected Forest, Southwest Sumba, work on the *kaliwu* agricultural system. This is further supported by community initiative or participation known as *lende ura* which is a local concept of community participation with a focus on all ecological components related to the function of forests as rain bridges. According to them, the preserved forest brings continuous rain, thereby, enabling the implementation of the *kaliwu* dryland mixed farming system. Meanwhile, the utilization of forest, water, and land resources is regulated through a customary agreement governed by *rotu*. This *Rotu* regulates the rights and obligations of each individual and also provides sanctions for violations and the public awareness for its compliance cannot be separated from the needs of the area which are presented as a ritual for the *marapu*, *marapu* forest, and water sources (Njurumana and Prasetyo 2010).

The landscape character of the traditional village is also influenced by the diversity of its vegetation us with some forest areas discovered to have customary or sacred forests (*kedawu erri*). Moreover, communities on the Sumba island generally use the selective cutting system to construct their traditional houses through the application of *kayu mayela* (*Artocarpus glaucus*), *kirru* (*Dyxoxylum caulostachyum*), *kunjul, kapali,* and *linnu* for the poles, *kayu langgapa* or *kayu pahit* (*Pierasma javanica*) and *kanunu* (*Drypetes ovalis*) for battens, *rotan* (*Calamus javencis*) for rope or fastening, and *alang-alang* (*Imperata cylindrica*) for the roof cover. The tectonics of these traditional houses show the forest does not only have an ecological meaning but also a spiritual meaning indicating the need to respect where

Marapu lives. Therefore, every tectonic element starting from its source to the placement needs to be handled carefully even through a ritual procession (Takandjandji 2015; Amin et al. 2016; Aziz et al. 2018; Gunawan et al. 2018; Dangga and Seran 2020).

The traditional communities in different tribes in Indonesia including the people of Sumba also use plants for several activities other than food and this is evident in their use for medicinal purposes, weaving crafts, and ritual ceremonies. The weaving craft is, however, part of their megalithic culture as observed with the use of hundreds of woven clothes to cover the bodies to be buried under a stone grave in the prehistoric era. Meanwhile, the dyeing process in woven fabric (*ikat*) production only uses plant materials grown in certain seasons and this is the reason it takes a very long time. The rainy season is for tying threads and preparing dyes, while the new coloring process starts in the dry season. It is important to note that the use of these natural dyes makes Sumba's woven products to be very distinctive and limited. The colors usually used are only red, black, and blue with the red produced from the root of *mengkudu* (*Morinda citrifolia*) mixed with *loba* leaves (*Symplocos sp*) as a color binder, while *Indigofera tinctoria* is the main ingredient for blue and black dyes. The additional plants usually applied in the process include the *kawilu*/candlenut (*Aleurites moluccana*) for color enhancement and *walakiri/dadap* (*Erythrina sp*) for fabric softener. Moreover, the plants used for medicinal purposes are *alang-alang* (*Imperata cylindrical*), *kerinyu* (*Chromolaena odorata*), *kesambi* (*Schleichera oleosa*), *landep* (*Barleria prionitis*), *sirih* (*Piper bettle*), and *tempuk wiyang* (*Emilia sonchifolia*) (Takandjandji 2015; Amin et al. 2016; Samadara 2019; Aziz et al. 2018; Dangga and Seran 2020).

3.7 *Nyale* and *Pasola* Tradition in West Sumba Traditional Village

The people of West Sumba Traditional Village have an agrarian socio-cultural and socio-economic background and also have a tradition of welcoming and celebrating harvest based on the beliefs of the local community as observed in most agrarian communities in the equatorial tropics. This is evident in the *Nyale* and *Pasola* traditions, especially for the *Kodi* ethnic group, which represent the relationship between cultural behavior and environmental settings. They are usually celebrated like the traditional *Marapu* religious ceremony in the local religion of Sumba tribes usually held annually between February and March in turns by four villages of *Kodi, Lamboya, Wonokaka,* and *Gaura.* The purpose is to welcome the new year and harvest in *marapu* belief with the place and time observed to be indicating the implementation of the bioregion concept in their residences (Handini et al. 2016; Lero 2018; Uma et al. 2018).

The *Pasola* is a tradition of dexterity games such as war which involves throwing wooden javelins at each other and is usually celebrated by the Sumba tribes on the island of Sumba, West Sumba, East Nusa Tenggara. Its uniqueness is in the spear-throwing competition between two groups from the backs of horses that are running fast. The name *Pasola* comes from the word *"Sola"* which is a kind of wooden javelin thrown at each other while riding horses and an affix *"Pa"* which emphasizes the activity is a game. It is, however, important to note that *Pasola* is not just a game due to the meaningful values it present to the people of Sumba such as the religious value of *marapu* belief, brotherhood, and kinship over hostility. There are also the heroism and sportsmanship values, socio-cultural values in the form of cooperation to ensure successful annual ritual, and economic value indicated by the exchange of goods and visitation during the ritual (Handini et al. 2016; Lero 2018; Uma et al. 2018).

The *Pasola* tradition cannot be separated from the *Nyale* tradition and this is evident in the conduct of the *Nyale* traditional ceremony to show gratitude for the gifts received based on the *Marapu* belief first before the *Pasola*. The tradition is marked by the arrival of the harvest season and the abundance of sea worms (*nyale*) on the beach. The determination to start the ritual is also very unique due to the need to meet several conditions and the celebration of *Pasola* depends on the *Nyale* received by the community. It is important to note that *Nyale* is usually observed on the ocean waters, while *Pasola* is on a wide expanse of field. The two settings are different but both hold very strong local geographical and ecological values and can be used to form a thesis on the existence of a tradition strongly directed towards maintaining the ecological values of the landscape. Moreover, the abundance of *Nyale* (sea worms) and the existence of horses are unique biological resources in the region with *Nyale* worms discovered to be a type of *Filum annelida* living in the southern ocean waters of Lombok and Sumba islands. Meanwhile, the horses are racehorses native to Sumba, Indonesia, and known as the *sandalwood* horse in ecological terms. This means these components are also ecological resources in the *Marapu* Village of West Sumba. It is, however, also possible to determine the territorial dimensions and properties of the bioregion concept of the *Nyale-Pasola* tradition as observed for the traditional farming system (Uma et al. 2018; Lero 2018).

3.8 Conclusions

The diversity of the Archipelago Cultural Landscape requires a flexible and adaptive development framework based on the unique local characteristics of a particular region. It is, therefore, possible to use the bioregion concept as an approach to overcome these problems by finding a balance between the human needs for livelihoods and the potential of natural resources in their area. This balance is, however, determined based on ecological, economic, and social criteria and this shows the bioregion concept is in line with the values and spirit of conservation to ensure the future sustainability of the Archipelago Cultural Landscape.

The relationship between the cultural landscape phenomena in the traditional villages of West Sumba-East Nusa Tenggara and Wallacea Region as a bioregion was used to explain the integrated and comprehensive working mechanism of the bioregion concept. The findings showed the potential of the bioregion concept to become a framework for research and policy design in developing the Archipelago Cultural Landscape due to its role in the territorial context affecting spatial policies. This further indicates its ability to integrate (i) territorial and (ii) socio-institutional dimensions.

These dimensions were able to explain the strong connection between the local wisdom of the traditional village of West Sumba as a form of the archipelago cultural landscape and the Wallacea region as a bioregion context. Moreover, the traditional villages in West Sumba, West Nusa Tenggara, and Wallacea region with their ecosystem characteristics were used to discuss the regional context of the cultural landscape entities associated with territories, places, natural environments, and time cycles.

Acknowledgments The authors express their gratitude to the Ministry of Research and Technology and the National Research and Innovation Agency of the Republic of Indonesia for their support in the Leading Basic Research of Higher Education (PDUPT) entitled "Model of Sustainable Traditional Village Architectural Landscape Management in Southwest Sumba Regency, East Nusa Tenggara." The authors also appreciate the contribution of the student community in East Nusa Tenggara, Architecture Department, University of Merdeka Malang.

References

Adams RL (2007) The megalithic tradition of West Sumba, Indonesia: an ethnoarchaeological investigation of megalith construction [Dissertation]

Agustin AF (2019) Desain Lanskap Rumah Budaya Sumba Nusa Tenggara Timur [Skripsi]. Institut Pertanian Bogor

Amin JJA, Rifai MA, Purnomohadi N, Faisal B (2016) Mengenal Arsitektur Lansekap Nusantara, 1st edn. Pustaka Pelajar, Yogyakarta

Ardiyansyah P (2014) Pengelolaan Lansekap Budaya dalam Kerangka Warisan Dunia: Studi Kasus Management Plan Lansekap Budaya Provinsi Bali. J Konservasi Cagar Budaya Borobudur 8(2): 68–73

Aziz IR, Rahajeng AR, Susilo (2018, April) Peran Etnobotani Sebagai Upaya Konservasi Keanekaragaman Hayati Oleh Berbagai Suku di Indonesia. Seminar Nasional Megabiodiversitas Indonesia, Gowa, Makasar

Bashari H, Cryan M, Dwisasanti N, Hermansyah A (2014) Hotspot Keragaman Hayati Wallacea. Burung Indonesia. https://www.cepf.net/sites/default/files/wallacea_ecosystemsummary_bahasa_1.pdf

van Berkel DB, Verburg PH (2011) Sensitising rural policy: assessing spatial variation in rural development options for Europe. Land Use Policy 28(3):447–459. https://doi.org/10.1016/j.landusepol.2010.09.002

Bisjoe ARH (2015) Kawasan Wallacea dan Implikasinya Bagi Penelitian Integratif Lingkungan Hidup dan Kehutanan. Info Teknis Eboni 12(2):141–148

Dangga MH, Seran W (2020) Pemanfaatan Jenis-jenis Tumbuhan Berkhasiat Obat oleh Masyarakat Sekitar Kawasan Hutan Lindung Rokoraka. J Wana Lestari 2(1):15–22

Djawa AR (2014) Ritual Marapu di Masyarakat Sumba Timur. Avatara, e-J Pendidikan Sejarah 2(1):71–85

Evanoff R (2017) Bioregionism: a brief introduction and overview. Aoyama J Int Polit Econ Commun 99:55–65

Fredholm S (2017) Making sense of heritage planning in theory and practice experiences from Ghana and Sweden. University of Gothenburg, Gothenburg

Geria IM (2014) Kearifan Ekologis Kampung Megalitik Rindi Praiyawang, Sumba Timur. Forum Arkeologi 27(2):99–108

Gunawan Y, Gunawan U, Umbu JC (2018) Konsep tempat pada Arsitektur Sumba Barat. Studi Kasus: Arsitektur Desa Weelewo [Research report]. LPPM Universitas Katolik Parahyangan

Hadi A (2018) Kapasitas Kesiapan 3 Desa untuk Mempromosikan dan Mengembangkan Pariwisata Berkelanjutan di Sumba Barat Daya [Assessment report]. Institute for Research and Enpowerment (IRE)

Handini R (2019) Pola Permukiman Kampung Adat Anakalang: Keberlanjutan Budaya Megalitik di Sumba Tengah. Kalpataru, Majalah Arkeologi 26(2):33–44

Handini GIM, Simanjuntak T (2016) Pesona Budaya Sumba. Gadjah Mada University Press. http://repositori.kemdikbud.go.id/11868/1/EBook_BudayaSumba.pdf

Hariyanto AD, Asri A, Nurdiah EA, Tulistyantoro L (2012) Hubungan Ruang, Bentuk dan Makna pada Arsitektur Tradisional Sumba Barat (Research report). LPPM Universitas Kristen Petra

Hasanuddin H (2015) Kebudayaan Megalitik di Sulawesi Selatan dan Hubungannya dengan Asia Tenggara [Dissertation]. Universiti Sains Malaysia

Hidayatun MI, Prijotomo J, Rachmawati M (2014) Arsitektur Nusantara Sebagai Dasar Pembentuk Regionalisme Arsitektur Indonesia. Transformasi Nilai-Nilai Tradisional Dalam Arsitektur Masa Kini. In: Seminar Rumah Tradisional 2014, pp 1–9

Kusumawati L, Topan MA, Winandari MR, Sofian I (2007) Jejak Megalitik Arsitektur Tradisional Sumba. Graha Ilmu

Laksono PM, Anantasari E, Nandiswara OA (2019) Reproduksi Moda (Pertukaran) Pangan: Menyemai Daulat Hidup di Sumba Barat Daya. J Masyarakat Dan Budaya 21(3):341–353

Lero SB (2018) *Strategi Pemerintah Daerah dalam Mengatasi Konflik Budaya Pasola* [Tesis]. STPMD "APMD"

Lockyer J, Veveto JR (eds) (2015) Environmental anthropology engaging ecotopia: bioregionism, permaculture, and ecovillages, 1st edn. Berghahn Books

Messakh J (2014) Akulturasi yang Mengedepankan Lokalitas dalam Membentuk Identitas Arsitektur Nusa Tenggara Timur. E-J Grad Unpar 1(2):178–188

Miller K (1996) Balancing the scales: guidelines for increasing biodiversity's chances through bioregion management. World Resources Institute, Washington, DC

Mross JW (2000) Cultural and architectural transitions of Southwestern Sumba Island, Indonesia. In: ACSA 2000 international conference, Hong Kong, China. Texas Tech University, pp 260–265

Njurumana GN, Prasetyo BD (2010) Lende Ura, Sebuah Inisiatif Masyarakat Dalam Rehabilitasi Hutan Dan Lahan Di Sumba Barat Daya. J Analisis Kebijakan Kehutanan 7(2):97–110

Poerwoningsih D, Antariksa A, Leksono AS, Hasyim AW (2015) Implementing visual resource management to support green corridor planning. Ecol Environ Conserv 21(4):2153–2160. http://www.envirobiotechjournals.com/article_abstract.php?aid=6547&iid=203&jid=3

Poerwoningsih D, Santoso I, Wahjutami EL (2018) Konsep Bioregion dalam Pengelolaan Sumberdaya Lansekap Arsitektur Nusantara. Mintakat J Arsitektur 19(1):49–58. https://doi.org/10.26905/mj.v19i2.3219

Reny EK, Subroto TYW, Saifullah A (2018) Konsep Ka'bani–Mawinne dalam Arsitektur Rumah Tradisional Sumba di Kampung Tarung Sumba Barat. Tesa Arsitektur 16(2):94–104

Rheindt FE, Prawiradilaga DM, Ashari H, Suparno GCY, GWX L, Wu MY, NSR N (2020) A lost world in Wallacea: description of a montane archipelagic avifauna. Science 367(6474): 167–170. https://doi.org/10.1126/science.aax2146

Ryan JC (2012) Humanity's bioregion places: linking space, aesthetics, and the ethics of reinhabitation. Humanities 1(1):80–103. https://doi.org/10.3390/h1010080

Samadara S, Sir JS, Samadara PD (2019) Pemberdayaan Perempuan Pengrajin Tenun Ikat di Kampung Prai Ijing, Desa Tebar, Kecamatan Kota, Kabupaten Sumba Barat, Nusa Tenggara Timur untuk Meningkatkan Perekonomian Keluarga dan Mendukung Pengembangan Pariwisata Daerah. Jaka J Jurusan Akuntasi 3(2):20–29. https://doi.org/10.32511/jaka.v2i1.261

Santoso SA, Widiningtyas T(2019) Merajut Pelembagaan Pendanaan Konservasi Berkelanjutan Untuk Wallacea (vol 1). Yayasan Penabulu

Solihin L (2018) Indonesia, & Badan Pengembangan dan Pembinaan Bahasa. Arsitektur Tradisional Sumba di Kampung Adat Ratenggaro, Uma Mbatangu

Takandjandji M (2015) Strategi Pelestarian dan Pemanfaatan Biodiversitas Berbasis Masyarakat di Nusa Tenggara Timur, pp 184–192

Taylor K (2009) Cultural landscapes and Asia: reconciling international and southeast Asian regional values. Landsc Res 34(1):7–31. https://doi.org/10.1080/01426390802387513

Taylor K, Lennon JL (2011) Cultural landscapes: a bridge between culture and nature. Int J Herit Stud 17(6):537–554. https://doi.org/10.1080/13527258.2011.618246

Uma WKJ, Handayani D, Nurgiri YS (2018) Makna Nyale dalam Upacara Adat Pasola sebagai Upaya Pelestarian Budaya di Sumba Barat Nusa Tenggara Timur. HISTORIA J Program Studi Pendidikan Sejarah 6(2):347–364. https://doi.org/10.24127/hj.v6i2.1430

Vel J (2010) Ekonomi-uma: Penerapan adat dalam dinamika ekonomi berbasis kekerabatan (Ed. 1). HuMa

Wahl DC (2018) Bioregionism-living with a sense of place at the appropriate scale for self-reliance. Towards Life-Knowledge. https://bsahely.com/2018/07/27/bioregionism%e2%80%8a-%e2%80%8aliving-with-a-sense-of-place-at-the-appropriate-scale-for-self-reliance-by-daniel-christian-wahl/

Zasada I, Häfner K, Schaller L, van Zanten BT, Lefebvre M, Malak-Rawlikowska A, Nikolov D, Rodríguez-Entrena M, Manrique R, Ungaro F, Zavalloni M, Delattre L, Piorr A, Kantelhardt J, Verburg PH, Viaggi D (2017) A conceptual model to integrate the regional context in landscape policy, management and contribution to rural development: literature review and European case study evidence. Geoforum 82:1–12. https://doi.org/10.1016/j.geoforum.2017.03.012

The Importance of Biocultural Landscape Concept in Managing Protected Area: A Case of Tasek Bera, Pahang State, Malaysian Peninsular

4

Zakaria Hashim and Saiful Arif Abdullah

Abstract

Protected area has been recognized as part of conservation strategies managed by various agencies either at the state, national level, and non-governmental organizations. Generally, the concept of protected areas is not only focused on conservation and protection of biological diversity, but also involves marine ecosystems, historic sites, and unique cultures. Protected areas are crucial for the well-being of forest communities. Pristine lands help conserve traditional customs which often take place in the heart of the forests while meeting the vital needs of the communities. Protected areas play a vital role to ensure the sustainability of the area. Meanwhile, the concept of biocultural landscape focusing on the relationship between humankind and nature, and natural and cultural landscapes has often been given less priority in managing protected areas. Thus, this chapter highlights the importance of biocultural landscape values and characteristics in protected areas management. Being one of the protected areas, Tasek Bera RAMSAR Site located in Pahang State of Malaysian peninsular has been selected to explore the importance of biocultural landscape concept to the management and sustainability of the protected area.

Z. Hashim (✉)
Faculty of Architecture, Planning & Surveying, Centre of Studies for Parks & Amenity Management, Universiti Teknologi MARA, Shah Alam, Selangor Darul Ehsan, Malaysia
e-mail: zakaria_hashim@uitm.edu.my

S. A. Abdullah
Institute for Environment and Development (LESTARI), Universiti Kebangsaan Malaysia, Bangi, Selangor Darul Ehsan, Malaysia
e-mail: saiful@ukm.edu.my

S. A. Abdullah et al. (eds.), *Conserving Biocultural Landscapes in Malaysia and Indonesia for Sustainable Development*,
https://doi.org/10.1007/978-981-16-7243-9_4

Keywords

Biocultural landscape · Biocultural diversity · Malaysia · 'Orang asli' Semelai · Protected area management · Tasek Bera RAMSAR Site · Traditional value

4.1 Introduction

Protected areas (PAs) are part of the foundation for global conservation administered and managed by various national, state, non-governmental organizations (NGOs), and possibly individuals in the case of private protected areas. They are the basis of almost all nationalities and international conservation strategies, set aside to be maintained serves as a natural ecosystem, acting as a shelter for the species and sustaining ecological processes which most strictly managed (Dudley et al. 2013; Macgregor and van Dijk 2014). The concept of protected area generally not restricted to the conservation and protection of flora and fauna but also involves marine areas, sites historic and culture which has characteristics that unique to certain community (Lockwood 2010; Lockwood et al. 2008).

Malaysia is quite fortunate to have diversity of flora and fauna in which makes it a unique country not only known as one of the world's biological mega-rich countries but with flora and fauna only available in certain areas. The country's tropical forests are estimated to contain 15,000 vascular plants (MNRE 2016). Fauna diversity includes 307 known species of mammals, 30 of which are endemic to Malaysia, 785 species of birds, 242 species of amphibians, and 567 species of reptiles, as well as 2068 species of freshwater and marine fishes (MNRE 2016). The possibility of more information stored in the tropical forest in this country is still unknown and of course, this should be explored. It is biodiversity richness that constitutes an extraordinary natural landscape that maintains our natural environment and the life-support systems that give us food, water, and numerous economic benefits. We depend on it for our very existence. Thus, the establishment of protected areas has been introduced to safeguarding our biodiversity (IUCN 2008). Meanwhile, conservation and sustainable development are two prominent words in the world today. The former has years of evolvement from fortress conservation to neoliberal conservation (Brown 2002), while the latter promotes the idea of sustainability in a global world. The link between conservation and sustainable development is notable enough in PAs management. In these natural areas, interactions occur among several factors: local communities, protected area managers, NGOs, politicians, and private ecotourism operators.

Malaysia is a megadiverse country, and it has ratified the Convention on Biological Diversity (CBD) in 1994. Since then, Malaysia has been promoting biodiversity conservation as an integral part of sustainable development which later became the policy theme for the Seventh and Eighth Malaysia Plans in 1996–2000 and 2001–2005, respectively. Within that period, a national policy on biodiversity conservation has also been adopted. Malaysia advocates a development path that emphasizes conservation while maintaining economic development.

Despite advocating such a development strategy, the challenges ahead for conserving and protecting the total biodiversity through the establishment of PAs remain serious. However, Malaysia, like any other countries, is also experiencing challenges in managing its biological diversity. In the process of the country pursuing to become a developed high-income nation, it also provides challenges including threats to its biodiversity. Other pressures that threaten Malaysia's biodiversity include habitat fragmentation, invasive alien species, pollution, poaching, increasing competition for land as well as climate change (GBO 2020). Furthermore, there is a general lack of awareness on the importance of biodiversity throughout the country as well as significant knowledge gaps. Undoubtedly, economic growth is essential for the well-being of the community. However, it also puts on the pressure to the flora and fauna. Most of the plants and wild animals are at risk. No doubt, conservation efforts through protected areas are important for the protection of biological diversity. Biological diversity or commonly known as biodiversity is important for environmental and economic resources which support the natural system and improve human welfare. While biodiversity is important to the country, the existence of the local community (i.e. neighbours) to the protected areas also cannot be denied.

Conversely, issues on biodiversity conservation have become one of the prime issues in managing natural landscape. The importance attached to the biodiversity has changed the whole scenario of the natural resource's values and their development potential. The ever-increasing pressure for biodiversity conservation has influenced the way we managed our natural resources. The need to reduce deforestation and forest degradation and to protect, conserve, and ensure the sustainable use of genetic resources remained a widespread concern requiring actions. Therefore, if we do not manage these challenges carefully and through a good strategy, our natural environment, life-support system, and economic stability will be affected (MNRE 2016).

Thus, the management of a protected area should not only focus on the conservation of biological diversity but also need to strengthen it management by considering the importance of biocultural landscape of the area. Indirectly, this will help the sustainability of management in the protected areas. Hence, this chapter will elaborate on the importance of the biocultural landscape concept in protected areas management. In this regard, Tasek Bera RAMSAR Site, one of the protected area located in Pahang state of Malaysian peninsular has been selected as a case study.

4.2 Tasek Bera RAMSAR Site

Tasek Bera (*Bera Lake*) is the largest freshwater lake, approximately about 25 km from north to south in the middle of Malaysian peninsular in the state of Pahang (Fig. 4.1). The main habitats are open water, reed/sedge swamps, and lowland forests. The uniqueness of it stems from the fact that the lake drains northwards through Bera river which then goes into the longest river in Malaysian peninsular, Pahang river.

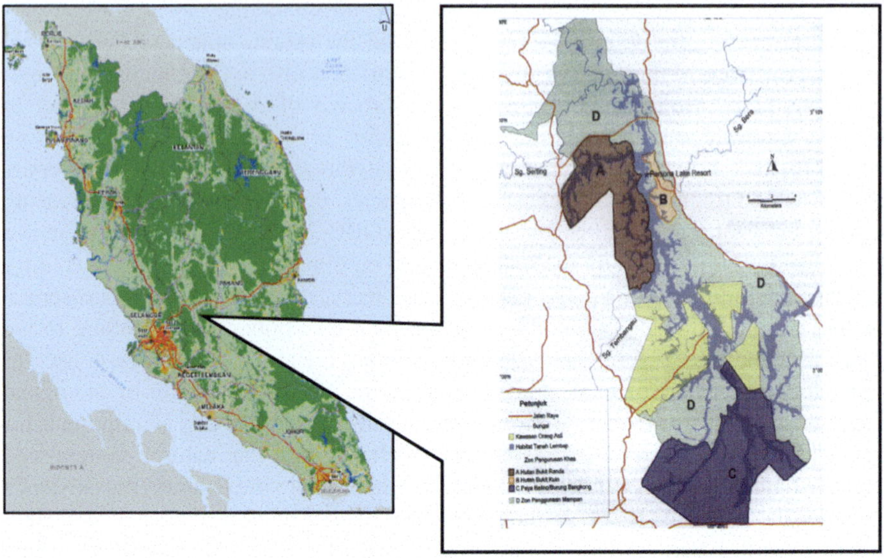

Fig. 4.1 The location of Tasek Bera RAMSAR Site in Malaysian peninsular (Source: WIAP 1999)

The PA of Tasek Bera is an area of 31,255 and 6800 ha of these is wetlands (DWNP 2013). This area was a dense forest 40 years ago, which can only be accessed from outside with great difficulties. Based on its biological diversity and value of ecological and socio-economic potential (Table 4.1), Tasek Bera was designated as the country's first wetland of international importance, or RAMSAR site in 1994. Subsequently, a Management Plan (MP) was established with the main goal to promote ecological conservation and sustainable use of natural resources. In addition, Tasek Bera was gazetted as Permanent Forest Reserve on 30th August 2007 under the National Forestry Act 1984. The forest surrounding the lake is home to Semelai people, one of 'orang asli' tribe in Malaysian peninsular (Fig. 4.2).

4.3 'Orang Asli' Semelai

In Malaysian peninsular indigenous people are known as 'orang Asli' which literally means 'original people'. The Semelai are indigenous people from Malay ethnic group (*Proto Malay*). The distribution of Semelai community can be found in central Pahang such as in Tasek Bera, Sg. Bera, Sungai Teriang, Paya Besar, and Paya Badak as well as at the Pahang border leading to state of Negeri Sembilan such as Sg. Serting, Sg. Lui, and Ulu Muar. The Semelai have been present at Tasek Bera for at least 600 years (Kangayatkarasu 2003) and during the insurgency period, they were resettled at Pos Iskandar. According to Elagupillay et al. (2008), the Semelais consider Tasek Bera as their sacred home. They are still dependent on it for their

Table 4.1 The characteristics of Tasek Bera RAMSAR site

Characteristics	Contents
Hydrological functions	• Purification of water
Carbon sink	• Flood control and flow regulation
	• Organic matter stored in the form of peat
	• Alteration of the ecosystem will result in the release of carbon dioxide, thus increasing the temperature
Importance for biological diversity	• 374 plant species
	• 94 fish species
	• 230 bird species
	• 68 mammal species
Occurrence of endangered species	• Asian elephant
	• Malayan tiger
	• Tapir
	• Asian Arowana
Occurrence of endemic species	• *Cryptocoryne purpurea* (an aquatic plant)
Semelai's community	• Livelihood • Culture
Potential for recreation and tourism	• High level of biodiversity offers a range of activities for ecotourism
Potential for nature education	• Nature-based surrounding suitable for awareness programmes

Source: WIAP (1999)

Fig. 4.2 Tasek Bera is surrounded by various types of forests which is home to 'orang asli' Semelai (photo by Z Hashim)

livelihood and cultural purposes despite the shift towards growing crop plantations due to exposure to mainstream conditions (Kangayatkarasu 2003). There are reportedly more than 3373 of the Semelai population at Tasek Bera (JKOA 2021).

The largest community of Semelai people in Tasek Bera is found at Pos Iskandar which consists of five main villages, namely Kampung Putat, Kampung Baapak, Kampung Gau, Kampung Jelawat, and Kampung Lengut or Kuin. Most of the Semelai lived along the banks and also deep in the forest surrounding the Tasek Bera. The Semelai people today live in a wooden thatch hut located in the villages. They are heavily dependent on the natural resources existed in the lake and its surround to farm, hunt, and collect forest products. As a native community in the area, they have a strong relationship with the land and freshwater lake as well as the vegetation surrounding it. The belief of the Semelai people that each hole of depth in the lake has a guardian has resulted in the lake and their land being preserved. Although they are still swidden practitioners, the Semelai conducted several eco-tourism activities to increase their socio-economic status. It is not surprising that the Semelai community gives names to each depth of the lake with the name of god as the owner. They strongly believe that each has its own guardian, among these are *jrolor* (big hand of the lake), *solok* (small hand), *padar* (swamp), and *ranoq* (swamp forest) (Azima et al. 2015). This clearly indicates the culture and survival spirit of Semelais is dependent on the existence of the lake itself.

They earn a living by tapping rubber from the six hectares of land sponsored by the government. Some of them work in palm oil plantations that dominate the surrounding of Pos Iskandar. Many of them are also wage earners who are employed in the nearby rural towns like Bahau, Triang, or Serting in Pahang. The natural environment is the source of their food, crafts, medicine, construction materials, recreation, cultural heritage, history, and identity. Significant anthropological research on the Semelai people was conducted by Hood Salleh (1978) and Gianno (1990). While economic studies of the utilization of wetland and forest products were carried out at Tasek Bera by Mohd Shahwahid (1997). This clearly indicates that their long-established heritage and culture had adapted to the forest environment. In other words, Tasek Bera undoubtedly represents features of biocultural landscape. The dependency of the local community particularly the Semelai to Tasek Bera is very high and they also have traditional methods, in the management of its resources.

4.4 The Biocultural Landscape Concept in Protected Area Management

The concept of biocultural landscape is centred on the relationship between humanity and nature, and natural and cultural landscapes (Hong et al. 2014). Biocultural landscape can be defined as the characteristics of a space that acts as a buffer so that biocultural diversity may be developed in a sustainable manner, ecological knowledge utilizing diversity and dynamic of the space mechanism to maintain the landscape development process, and an ecosystem which has an influence on the space (Hong 2007). The concept of biocultural landscape is an approach to heritage conservation closely related to biological diversity as well as cultural heritage inherited by the indigenous and local communities in a landscape area. In other words, biocultural landscape is an interaction of natural and human system coexisted. It is an essential look at a broader perspective rather than focusing on biodiversity without considering the importance of cultural diversity. Whereby environment has a close relationship with humankind and their culture. According to Hong (2014), biocultural diversity is important to conserve environment and human well-being. It is undeniable human depending on nature and cannot exist without utilizing natural resources.

Most of the studies usually have been restricted to studying traditional ecological knowledge of indigenous groups and their roles in nature conservation in developing countries particularly in Latin America, Asia, and South Africa (Pretty et al. 2009). In reference to the conservation approach through protected areas, it is often focused on the management of biological diversity which emphasizes on protection of species, habitats, and ecosystems. However, human involvement should also be emphasized to provide knowledge and environmental issues related to biodiversity loss. In this context, there is a close relationship between the existence of biological diversity, cultural values, and the dependency of human well-being or better known as the concept of biocultural landscape to ensure conservation management achieves its goals and in turn, ensures the sustainability of natural resource management in Malaysia particularly. Many researchers have indicated the importance of mutual interactions between biodiversity, nature, culture, and heritage (e.g. Agnoletti 2006, 2007; Agnoletti and Rotherham 2015). According to Hong (2007), there is significant interaction between nature and human creates culture through distribution of biological organisms and human's utilization. At the same time, ecosystems and habitats will suffer from destruction due to excessive human intervention through development and uncontrolled use of natural resources.

The direct relationships of those elements will certainly enhance the effectiveness of the management of protected area in this country. If the human intervention is not being monitored and managed properly it will affect our biodiversity resources. Therefore, it is necessary for stakeholders to come together beyond the existing concept of biodiversity. Diversity of landscape that had created biological and cultural diversity, however, is influencing both human tradition and life. However, one of the prevalent issues in the field of protected area management would be the relationship between protected area and its neighbours. This sweet–bitter relationship is loaded with protected area management theories and principles adopted by most management agencies mixed with values and norms of local community who often reside within or in close proximate to the protected areas (Daim 2015). The relationship between protected areas and its neighbours is often considered as a conflict rather than a good synergy. The mutual need of human and nature on the same landscape has changed over the years through development, technologies, knowledge, and socio-economic of the community.

4.5 The Importance of Biocultural Landscape Concept in the Management of Tasek Bera RAMSAR Site

In 2001, Tasik Bera RAMSAR Site Management Unit has been established responsible for managing Tasek Bera. This management unit is a collaboration between various agencies such as Department of Wildlife and National Park (DWNP) together with other agencies such as Department of Forestry, Department of Fishery, Department of Environment, and Department of Orang Asli Development (JKOA). Collaboration among various agencies is a good management approach. However, due to the different's views by the respective agencies in terms of task and enforcement functions cause the area to be poorly managed. Apart from that, Tasek Bera also facing large scale agricultural activities especially oil palm and other human encroachment. The expansion of oil palm plantations indirectly affected the Semelai people, jeopardizing the original resources of the wetland area and insufficiently provides the tribe. Waste disposal and chemicals from agriculture activities worsened the situation and affected the surrounding population. The destruction of the Tasek Bera has also been reported by Chong (2007), concerning the deterioration of water quality and quantity.

Thus, it is vital to all the stakeholders in Tasek Bera to plan and implement sustainable and conservation management practices. Indeed, this is in line with what has been suggested in the concept of biocultural landscape which taking into account the importance of heritage, culture, and interaction of human with the biological diversity of the areas (Agnoletti and Rotherham 2015). The understanding of biocultural landscape among the stakeholders is essential to ensure the efficiency of management of this wetland area. There are two methods of management practiced by responsible parties, namely with an active and passive community participation approach. It is aimed at establishing a healthy management condition to manage a protected area. The co-management concept is often used and adapted within the management of protected areas around the world. According to Local Government New Zealand (2007), co-management can be defined as decision-making processes where more than one party is involved in the process. This is where adaptation of the importance of biocultural landscape can be applied where locals are allowed to preserve their heritage conservation related to biological diversity as well as cultural heritage inherited by the indigenous and local communities in a landscape area.

Considering the Semelai is the main stakeholders hence the management needs to involve them in any consultations, cooperation, and communication. According to Haidar (2014), the co-management practice has been adopted through various activities and programmes especially with the Semelai where they participate as part of committee members. The management of Tasek Bera is often considered the views and opinions of the Semelai as important to improve the management of the area. The management and Semelai have participated in communal work programmes and the management has identified some of the additional activities such as ecotourism and aquaculture training programmes for the Semelais to improve their livelihood. However, the extent of the Semelais involvement in the management decision-making remains to be seen. Community conservation, currently touted as a possible solution to the evils and ills of conventional modes of nature preservation, is being formalized in ways that threaten biocultural diversity. Having managed their forests communally according to customary practices for generations, the community has a rich and deep cultural and ecological knowledge.

4.6 Discussion and Conclusion

Indeed, it is clearly stated, as a wetland and protected area Tasek Bera is not only referring to the biological diversity alone but also includes the heritage and cultural values that exist in a landscape area required to be conserved and protected. In the case of Tasek Bera, the wetland is a home for abundance of biodiversity. While ecosystems do not only consist of physical attributes but they are also subjected to and influenced by cultural perceptions and values (Verschuuren et al. 2010). By looking at the characteristics of Tasek Bera, the Semelai community has a very close relationship with the natural environment and their cultural and heritage elements need to be considered in conserving the area. According to Agnoletti and Rotherham (2015), historical rural landscapes are often related to traditional practices and provide important for understanding biocultural diversity. It is true that in some cultures, the spiritual values of certain sites and species may be important enough for local people to conserve and protect the ecosystems that contain them (Verschuuren et al. 2010).

The lake and the nature surrounding the area have a spiritual significance which led to the culture and heritage of the Semelai. According to Dudley (2008), sacred sites and species are increasingly known for their significant contribution to biodiversity values. In the context of protected area management, sacred sites and heritage with the historical and cultural value of the local community play a vital role in understanding the interaction of humankind and nature (Dudley and Stolton 2008). On many occasions, such cultural perceptions have proved to be invaluable in building approaches to sustainable management. Thus, information on biocultural landscape also plays a vital role in developing protected area management categories. Significantly, all the protected area management categories (Ia, 1b, II, III, IV, V, and VI) not only focus on the primary objectives, distinguishing features but also the important role of its landscape or seascape elements (IUCN 2008).

Based on the classification of protected area management category, Tasek Bera RAMSAR Site has been classified as Category V (DWNP-DANCED 1996). Notably, this category is defined as a protected area where the interaction of people and nature over time has produced an area of distinct character with significant ecological, biological, cultural, and scenic value. Conversely, safeguarding the integrity of this interaction is vital to protecting and sustaining the area and its associated nature conservation and other values (Dudley 2008). Thus, there is an opportunity for the authorities to cooperate with local communities to develop a community-based conservation practice in this area. There is a close relationship between protected areas, the local community, and the natural setting. Some of the flora and fauna are sources of food and cultural importance to the Semelai. Upon the establishment of Tasek Bera as a RAMSAR site, most of the Semelai face an issue that led to conflict with the management of Tasek Bera (Haidar 2014).

It is highly hoped that the active participation of the Semelai in decision-making can be improved in the future by the existing management. Active participation can contribute to effective stakeholder management, biodiversity, and heritage conservation. Indigenous peoples ('orang asli' in Malaysia) and local communities manage over half of the world landscapes (Franco 2021). Although there is an initiative to include the indigenous in the management and programme organized by the respective authorities in managing the protected areas; however, their involvement is still considered at a low level. It is very important to integrate all the information on biocultural landscape elements which will strengthen overall biodiversity conservation and also the sustainability of the protected area.

References

Agnoletti M (ed) (2006) The conservation of cultural landscapes. CAB International, Wallingford, p 320

Agnoletti M (2007) Guidelines for the implementation of social and cultural values in sustainable forest management. A scientific contribution to the implementation of MCPFE—Vienna Resolution 3. IUFRO Occasional Paper No. 19, Vienna

Agnoletti M, Rotherham ID (2015) Landscape and biocultural diversity. Biodivers Conserv 24: 3155–3165

Azima AM, Sharifah Mastura SA, Saad S, Awang AH (2015) Orang Asli Semelai: conflict of defending land ownership rights. Mediterr J Soc Sci 6(4 S3):63. https://doi.org/10.5901/mjss.2015.v6n4s3p63

Brown K (2002) Innovations for conservation and development. Geogr J 168:6–17

Chong G (2007 Asek Bera: past, present and future. Paper presented at the colloquium on lakes and reservoir management, status and issues, 2–3 August 2017. Putrajaya, Malaysia

Daim MS (2015) Community participation in protected area management in peninsular Malaysia: a case study of Taman Negera national Park, Pahang, Malaysia. Unpublished Thesis. UiTM

Dudley (2008) Guidelines for applying protected area management categories. IUCN, Gland

Dudley N, Stolton S (2008) Defining protected areas: an international conference in Almeria, Spain. IUCN, Gland, 220 pp

Dudley N, Stolton S, Shadie P (2013) Guidelines for applying protected area management categories including IUCN WCPA best practice guidance on recognising protected areas and assigning management categories and governance types. IUCN

DWNP (2013) Compendium conserve & care. Marza Excellent Enterprise, Selangor

DWNP-DANCED (1996) A master plan on capacity building & strengthening of the protected area system in peninsular Malaysia. Department of Wildlife and National Parks, Kuala Lumpur

Elagupillay ST, Mokbolhassan HK, Kassim H (2008) Tasek Bera: the co-management model employed to administer the first Malaysia Ramsar Site, Lake Governance Workshop, Putrajaya, Malaysia, 19 August 2008. Academy Sciences Malaysia

Franco MJ (2021) A framework for identifying and integrating sociocultural and environmental elements of indigenous people's and local communities' landscape transformation. Persp Ecol Conserv 19(2):143–152

GBO (2020) Global Biodiversity Outlook 5. Secretariat of the Convention on Biological Diversity, Canada

Gianno R (1990) Semelai culture and resin technology. The Connecticut Academy of Arts and Sciences, New Haven, 238pp

Haidar KM (2014) Managing Tasek Bera: Department of Wildlife and National Parks and local community participation. J Wildl Parks 27:121–127

Hong SK (2007) Linking man and nature landscape systems: landscaping blue-green network. In: Hong SK et al (eds) Landscape ecological applications in man-influenced areas: linking man and nature systems. Springer, Dordrecht, pp 505–523

Hong SK et al (2014) Biocultural landscapes. Springer, Dordrecht

IUCN (2008) Defining protected areas. Gland, Switzerland

JKOA (2021) Jabatan Kemajuan Orang Asli. https://www.jakoa.gov.my/ (Accessed on 26 June 2021)

Kangayatkarasu N (2003) Biodiversity conservation in action-a case of Tasek Bera Pahang. Masters Thesis, University Of Nottingham

Local Government New Zealand (2007) Co-management: case studies involving local authorities and Maori. Local Government New Zealand

Lockwood M (2010) Good governance for terrestrial protected areas: a framework, principles and performance outcomes. J Environ Manage 91(3):754–766

Lockwood M, Worboys G, Kothari A (2008) Managing protected areas: a global guide. Ann Tour Res 35(1):278–279

Macgregor NA, van Dijk N (2014) Adaptation in practice: how managers of nature conservation areas in eastern England are responding to climate change. Environ Manag 54(4):700–719

Ministry of Natural Resource Environment (2016) National policy on biological diversity 2016-2025. Putrajaya, Malaysia

Mohd Shahwahid HO (1997) Economics of fishing, natural product exploitation and shifting cultivation at Tasek Bera. Integrated management of Tasek Bera Technical Report Series. Wetlands International—Asia Pacific, Kuala Lumpur

Pretty J et al (2009) The intersection of biological diversity and cultural diversity: towards integration. Conserv Soc 7(2):100–112

Salleh H (1978) Semelai rituals of curing, Unpubl. D.Phil thesis submitted to St Catherine's College, University of Oxford, 446pp

Verschuuren B, Wild R, McNeeley J et al (eds) (2010) Sacred natural sites, conserving nature and culture. EarthScan, London

Wetlands International-Asia Pacific (1999) Integrated management of Tasek Bera—summary of Tasek Bera management plan. Wetlands International-Asia Pacific, Kuala Lumpur

Bridge and Islandness: The Case of Suramadu Bridge in Indonesia

5

Sun-Kee Hong, Luchman Hakim, and Argus Faqot

Abstract

The natural resources, the nature of an island's ecosystem, and the minds of the islanders who live there determine an island's identity. It is thought that the views and perspectives of people on the land, people who look at the island as an object, determine the islandness of a specific island as well. Natural phenomena, including the physical, geographical, and ecological characteristics of the island, which have been associated with the biological culture of the islanders for a long time, also determine its islandness. However, old island traditions, perceptions and identities are changing due to a number of recent developments. The existence of bridges connecting islands to other islands and islands to the mainland significantly changes an island's identity. Products such as vegetables and fruits harvested from the islands can be quickly delivered to the mainland, while people can travel to and from the island in their vehicles at any time. They no longer have to worry about ferries being delayed or cancelled due to bad weather. Conversely, a number of studies have shown that there have been negative impacts since the construction of bridges connecting the mainland to islands. Pulau Madura in Indonesia is a large island with a population of more than 3 million people. It was connected to Surabaya, Indonesia's second largest city, by a bridge in 2009. This chapter briefly introduces the results of a survey on the changes in perception and

S.-K. Hong (✉)
Institution for Marine and Island Cultures and Department of Liberal Arts, Mokpo National University, Mokpo Campus, Songrim-ro, Jeollanamdo, Republic of Korea

L. Hakim · A. Faqot
Department of Biology, Faculty of Mathematics and Natural Sciences, Brawijaya University, Malang, East Java, Indonesia
e-mail: luchman@ub.ac.id

S. A. Abdullah et al. (eds.), *Conserving Biocultural Landscapes in Malaysia and Indonesia for Sustainable Development*,
https://doi.org/10.1007/978-981-16-7243-9_5

opinions of Pulau Madura residents on their connection to the mainland with the construction of the Suramadu Bridge. In terms of economic activity, most of the natives of Pulau Madura gave a positive assessment. Some gave negative feedback because of the increased motorcycle traffic accidents and the change in identity that Madura youths are undergoing due to the influx of western culture from metropolitan Surabaya. Surabayan citizens, however, were not very interested in the bridge. According to them, there are few interesting places to see on Pulau Madura, which means there are not many tourist destinations. In their opinion, it is necessary to develop some tourist destinations on the island to make it more attractive.

Keywords

Archipelago · Biocultural diversity · Indonesia · islandness · mind · Pulau Madura · Suramudu Bridge · Traditional knowledge

5.1 Islandness as an Indicator of Biocultural Diversity

There are expressions such as "good-natured" and "good-hearted," which show that personality comes from the heart. Nature is a natural heart (the so-called mind) and is construed as the true and unchanging essence of mind. A number of clans, from those who arrived on islands many centuries ago to those who settled there more recently, form various villages and live together on both large and small inhabited islands. In most cases, the clans who settled on the island first form mainstream island society. In general, those who arrived later set up their base in other (uninhabited) places. In other words, there are clan-specific areas on islands. These areas are characterized by the history of settlement and the nature of the minds of the inhabitants. If the early arrivals were amenable, they would live together with the next arrivals as neighbors. We have not yet seen an island with only one surname. If there are any, it would be a very interesting topic for research on islandness (Hatfield and Hong 2019). Some people say that the island is closed, while others contend that it is open because it is open to the sea. Closedness and openness are two-sided characteristics of the island, and two-sidedness itself is islandness. The island is both closed and open because of its unique geographical characteristics. In particular, such islandness is differentiated by what surrounds the island (Conkling 2007).

The criteria for defining islands, such as large and small islands, islands adjacent to land, islands within an archipelago, vary greatly depending on the island, sea, and geopolitical role (Hong et al. 2014b). However, directly or indirectly, the minds of islanders affect islandness itself. In other words, islandness is closely related to the identity and sociality of the people who first set foot on the island. As the saying "the heart of the village" goes, each island has a different heart and emotion. Just as each person has different genes, island-like features, that is, the characteristics of islandness are well-established and differentiated depending on the combination of people that make up an island's society (Hong et al. 2014a, 2018).

Islandness is sometimes decided consciously by the people living on the island. In other words, they use the island to express their nature as they perceive it in their minds. The description of an island, mainly expressed in creative works such as novels or poems, also plays a role in anchoring its specific islandness. For instance, a writer can describe the innermost sanctum of an island as if he or she had lived there for a long time, even though this was not the case. So, if you were to say "You've never lived on this island!," there would be many writers who could not deny it.

It would not be correct to see the closedness and openness of the island separately from a dichotomous perspective. Islandness is not simply determined by geographical features or spatial accessibility, but also by the nature of the residents on the island. Islanders are not unsociable because the island is closed, and islanders are not open-hearted because the island is open. The island is multifaceted in nature and color. You can say that it is like a kaleidoscope. The minds of the islanders then serve as an enzyme for ripening this multifaceted islandness. When you visit and tour various islands around the world, you feel that each island has a different color. Although some islands are adjacent to each other in terms of distance, they often have different cultures and customs. Through the construction of a bridge connecting the island to the island and the island to the city, an island's unique biocultural diversity will change and urbanization will proceed. The construction of bridges between Japan's Seto Inland Sea, Korea's many islands in the archipelago, and Indonesia's Surabaya-Madura Island promotes social, economic, and cultural change (Effendi and Hendarto 2014; Hong 2011, 2015; Hong and Nakagoshi 2017; Izzah et al. 2021). The region will become a daily life zone (a living space of a day) through bridges that connect adjacent islands, and the islands to the mainland. Despite this, an island remains an island. Each island contains a mix of the resident clans' various cultures, including those of the initial ancestral clans and their villages. Over such a long time, the genetic mix of an island's society forms its islandness. We wonder if these island-specific features, that is, the islandness, will disappear due to bridge projects.

5.2 Bioculture: Ecological Knowledge and Information Between One Island and Another Island

Recently, islands have been experiencing changes in their marine environment due to the climate crisis, social changes resulting from aging and depopulation, and cultural changes caused by increased accessibility (Hong and Nakagoshi 2017). Islands around the world, including in Indonesia, are undergoing changes in identity due to internal and external pressures. The scope of these changes is amplified by accessibility and tourism policies which are advanced by mainland-to-island and island-to-island bridges, seawalls, and airports. Nevertheless, some islands have their own features that are not affected by internal and external pressures and which have a different sensitivity to the speed and frequency of change. Planning and implementing island policies should focus on the uniqueness and identity of each island. What an island needs to have is its identity as well as its physical properties. What is essential for designating and assessing the identity of an island is its islandness.

Knowledge is a unique cultural gene of an era and is passed down through the generations (Maffi and Woodley 2010; Lee 2020). Knowledge includes traditional cognitive systems that humans have adapted to use limited resources in a barren natural environment and such knowledge is passed on to future generations as a tool for future survival.

Harvesting and fishing, the typical activities for living in coastal and insular areas, contain traditional knowledge of ecological and cultural value (Hong 2010, 2011, 2015). Simple tools, harvesting and fishing techniques, social institutions, and indigenous knowledge have proved to be highly efficient survival resources that have maintained an ecological balance through minimal human intervention in natural processes. Given this fact, people and cultures must be included in the entire ecosystem (e.g., academia uses the terms "landscape" or "seascape" as the expanded meaning, rather than the ecosystem). An ecosystem approach to the entire system is required in order to identify the knowledge system of an insular area (Hong and Pungetti 2012). Moreover, inheriting these cultural factors in a creative manner, preparing for modern living and future environmental problems, and finding ways to harmoniously satisfy ecological conservation, economic needs, and cultural diversity are ways of transmitting and preserving island knowledge.

Traditional knowledge develops into future knowledge. Currently, international science and technology related to fisheries and aquaculture have reached a level that can save special ecosystems of local communities and be used in an eco-friendly and ecological way. Human, social and cultural conditions are established to activate traditional cultural elements, or to motivate island residents' autonomous organizations, government and non-governmental organizations, and to invigorate ecological fisheries and aquaculture. However, they were not manifested because of the modern cultural and economic drivers that lead actions and intentions toward the exploitation of resources. In that sense, it is time for fishermen and fishing communities to become more aware of the conservation of the marine ecosystem.

Food is an intellectual heritage created by the traditional knowledge of island residents, using the biological resources of the seascape and islands. Food ingredients obtained from the seascape and tidal flats, including fish, shellfish, seaweed, and salt, are sources of nutrients such as proteins, vitamins, and minerals that are essential to human survival. The possibility of eco-friendly and ecological development depends on how an economic system, in which such resources are acquired, distributed and consumed, is organized. Since ancient times, the archipelago has shown very remarkable fish diversity as it has many islands, a variety of topography, seawater flow channels, and geological foundations containing nutrients such as what is found in mudflats. In particular, in these areas where mudflats have developed, people have continued not only deep-sea fishing activities but also various activities in the tidal flats (Hong et al. 2014b). As we have previously suggested, we think it would be good to use the term "Mudflatter" to refer to them as there are native words such as mariner and islander (Fig. 5.1).

Fig. 5.1 A fishing village in Madura Island (Pulau Madura) where anchovies are dried. The seascape and tidal flats at low tide are very similar to those of Japan, China, and Korea (Photo by SK Hong)

Identifying the boundary of the sea, the topography and geography of the island and its biological resources is very important knowledge for island life (Conkling 2007). The peninsula is a passage for goods to the sea, and the island is a stepping stone reaching from the peninsula to the sea. As such, islands and peninsulas have been used as important corridors for goods and human communication and exchange. Salt, together with fish and seaweed, are the main products of an archipelago. Salt produced in the small islands of Indonesia is sun-dried salt, a product produced by seawater, mudflat soil, and solar heat, based on a production system that relies almost entirely on natural processes. Making salt is also a rice farming method. As the salt ripens, the rice becomes ripe. "Mudflatters" are masters of traditional ecological knowledge who collect seafood, produce salt, and grow rice from mudflats. They are able to read nature laterally.

5.3 Changes in Islandness of Pulau Madura

Pulau Madura is a large island off the northeastern coast of Java Island, Indonesia. It has an area of 5025 km^2 and a population of 3,720,000 as of 2014, making it one of the most populous islands in Indonesia. This island can be reached by means of a land-to-island bridge constructed an hour away from Surabaya, Indonesia's second largest city. Construction on the Suramadu Bridge, named after Surabaya and the

Fig. 5.2 Suramadu Bridge viewed from a fishing village in Surabaya City (Photo by SK Hong)

Madura Strait, started in 2003 with capital from China and it was completed in 2009 (Fig. 5.2). The length of the bridge is about 5438 m. The people of Pulau Madura are mostly Madurese ("Orang Madura" in Indonesian). Historically, Pulau Madura was a colony of the Javanese but the Madurese are very independent and have been able to maintain their livelihood. They are found not only in Indonesia but throughout the world (Whitten et al. 1996; Wekke et al. 2019). It is said that if one Madurese person moves somewhere to live, they will gradually increase in numbers and occupy the area in about 30–40 years.

The authors visited Pulau Madura and Surabaya city to conduct interviews with residents in January 2019. We wanted to survey the changes in residents' perceptions before and after the construction of the Suramadu Bridge. We visited and surveyed four major cities (Bangkalan, Sampang, Pamekasan, and Sumenep) according to distance from the bridge as shown in Fig. 5.3. In Surabaya City, we conducted our survey on residents of coastal villages near the bridge. The survey included questions about the changes before and after the construction of the bridge, the positives and negatives of the Suramadu Bridge, personal interests, and the bridge's relevance to industry. For reference, questions were given about major products in Pulau Madura in order to look into the possibility of developing an industry in relation with Surabaya City afterwards (Table 5.1).

Indonesia is divided into Sumatra, Java, Kalimantan, and the Lesser Sunda Islands. The biodiversity of these areas is different because their soil and vegetation vary depending on their geographical features. While Java has rich soil fertility, Pulau Madura is a limestone area and has soil with low fertility and thus has low crop productivity. Natural vegetation has adapted to dry environments, including teak (*Tectona grandis*) and numerous species of bamboo (Whitten et al. 1996; Kosim 2012). The Madurese have been left with only the toughest vegetation that can survive in the harsh island environment. In the past, access to Java and other

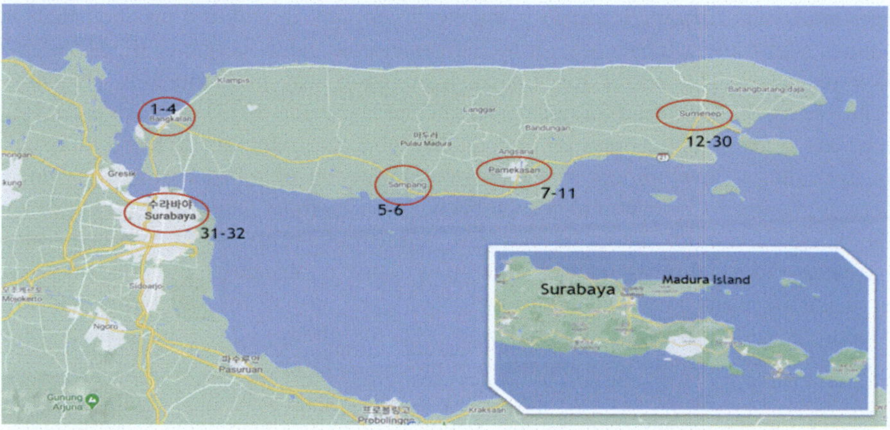

Fig. 5.3 The surveyed locations and investigator numbers in Pulau Madura and Surabaya City, Indonesia

Table 5.1 Questionnaire to survey the changes in the residents' perceptions before and after the construction of the Suramadu Bridge

No	Name	Ages	Occupation	Location	Questionarie				
					Positive impact	Negative impact	Personal perceptions	Relation with industry	Bio-resources

surrounding islands was by traditional boats and ferries from two main ports, Kamal in Bangkalan and Kalianget in Sumenep.

In 2003, the central government started to build the Suramadu Bridge to connect Surabaya and Madura. The objective was to stimulate development in Madura. The bridge was completed and opened in 2009. Once Pulau Madura was connected to Surabaya by bridge, a substantial chunk of the island's population migrated to Java. Nevertheless, the island is not depopulated because of the Madurese traditional customs of early marriage and fecundity. They usually marry between the age of 14 and 17 and have four or five children. However, according to the residents here, a lot of outsiders are expected to come in due to the bridge. It is believed that the unique and strong identity of the Madurese will gradually change.

The major industries of Pulau Madura include fishing and production of salt, corn, and other agricultural crops (Izzah et al. 2021). Salt farms produce salt during the dry season from June to September, and the rest of the time they are used as a shrimp aquaculture farms. Agriculture is mainly field farming. It is said that the soil of Pulau Madura has been barren since ancient times and corn is the only food crop that grows. This tradition continues nowadays. The Madurese plant corn in the rainy season and tobacco in the dry season to earn their income. The residents in Pulau Madura make the most of the seasons through inherited traditional knowledge about agriculture and fishing. Tobacco is another important agricultural product. Recent studies confirm that tobacco, with its high economic value, is one of the islanders' favorite crops. The center for tobacco cultivation was in Pamekasan Regency, followed by Sumenep and Sampang (Hasan and Darwanto 2013). Madura is considered one of the potential areas for high quality tobacco production. In line with issues of agricultural development, the intensive cultivation of tobacco recently gained serious attention. Compared to corn, tobacco provides more income which improves the socio-economic status of farmers. Numerous cultivars of tobacco were introduced to Madura Island as a strategy to improve tobacco production (Suwarso et al. 2020). Mangrove forests are well developed around the salt farms, and the Madurese catch, process, and send crabs that live there to Java. In this way, the residents of Madura are also interested in processed foods using salt and fish.

Bangkalan, a major city on Pulau Madura, was occupied by the Dutch East India Company in the seventeenth century, which led Chinese merchants to settle there. Dutch and Chinese-style buildings of the time stand in a continuous row on the island. They are falling into ruin through lack of management. We think that the existence of these buildings from various eras would be enough to make Bangkalan an island city worth listing as a UNESCO World Heritage Site. It was a thriving trade port in the past and there are many buildings with different architectural styles and various religious facilities that remain. There are various shrines dedicated to Mazu, a Chinese sea goddess, Catholic churches, Islamic mosques, and Buddhist temples. We think this place was a melting pot of diverse cultures. On our visit to fishing villages near the coast of Pulau Madura, we looked at the fish the residents had just caught and interviewed them about the impact of the land-to-island bridge.

Most residents are positive about the bridge, but it is a natural outcome that the economy and culture of Pulau Madura are changing internally. Some results showed that the culture of the Madurese, who by nature are clearly able to maintain their livelihood, will remain unchanged. However, they will become more united due to the construction of the bridge.

Residents of the Bangkalan region, close to the Suramadu Bridge, gave positive evaluations of the bridge as the connection to the mainland led to more active trade activities, increased tourism, and higher income for local residents (Table 5.2). However, they revealed the negative impact that the increase in the number of vehicles has had, leading to an increase in car accidents in the city and a rise in air pollution. Residents of the Sampang region in Pulau Madura were also relatively positive. They said that transportation has become easier with the bridge and positive perceptions of Pulau Madura have increased. Above all, it was noted that the efficiency of trade has increased due to the faster transportation speed on the bridge than on a ship resulting in cost savings. Conversely, there was a worry that the western culture of the metropolitan city, Surabaya, could disrupt the traditional culture of Pulau Madura, confirming that the residents of the island also thought critically about the island's identity. The Pamekasan and Sumenep regions of Pulau Madura, the most remote regions from the Suramadu Bridge, revealed a positive attitude to the increased trading activities and Muslim tourists, while they saw the moral change in the younger generation due to urbanization as a threatening factor

Table 5.2 A questionnaire on the perceptions of islanders

					Questionnaire				
No.	Name	Age	Occupation	Location	Positive impact	Negative impact	Personal perceptions	Relation with industry	Bio-resources
1	Local 1	67	Trader	Bangkalan	Happy with Suramadu bridge because it is able to increase income, especially for traders	Not feel bad impact	With the establishment of the bridge, it is expected many people will come and increase trading activity	Accessories from mollusk shell, accessories including SAKERA clothes accessory	Coconut, rice
					Distance between Madura and Java become close, compared to the past				
2	Local 2	60	Trader	Bangkalan	Bangkalan becomes big town/city since many people come and visit the Suramadu Bridge		Suramadu safe, without terrorist and vandal attacks. It is important to facilitate vehicle movement safety	Madura Herbal medicine (Jamu Madura UD. Ny. Nurul Zakiyah) Madurese Garment	Rice, corn, manioc, coconut
					Success of trading, especially among small and medium enterprises, increases income among local traders				
3	Local 3	32	Trader	Bangkalan	After opening Suramadi, I felt trading activity increased and I got more money.	Increase of car and motorbike potentially increase road accident in Bangkalan	Increasing safety of bridge	Tofu factory	Coconut, corn, rice, soybean

(continued)

Table 5.2 (continued)

No.	Name	Age	Occupation	Location	Questionnaire				
					Positive impact	Negative impact	Personal perceptions	Relation with industry	Bio-resources
					Access to Java easier				
4	Local 4	47	Trader	Bangkalan	Increase accessibility and people's mobility Surabaya-Madura. Increase tourist flow to Madura and therefore contribute to income	Increase of population	I wish all people pass the bridge safely	Madurese scarf	Rice, corn
19	Local 19	33	Lectures in university	Sumenep	Transportation become easy, cheap	Recent generation of Madura follows outside culture	Maintenance of bridge and road regularly	Crispy manioc, tofu and tempe	Manioc, soybean, corn, rice, tobacco
20	Local 20	22	Graduate student	Sumenep	Distance of Java and Bali closer	Ferry income decreased	We wish all people passing over the Suramadu to be safe	Keris, manioc crispy, crispy batik	Corn, manioc, rice
21	Local 21	26	Female housekeeper	Sumenep	Transport	Many local cultures under threat, decrease of local culture due to outside impact	Increase of Madurese prosperity in economic and culture and tourism	Tofu factory, tempe	Red onion, pepper, rice, soybean

Some survey data from Bangkalan and Sumenep regions, Pulau Madura (in this table, only 7 cases out of the 32 cases surveyed in Madura and Surabaya are presented)

Fig. 5.4 A row of shops built along the road of the Suramadu Bridge on Pulau Madura (Photo by SK Hong)

compared to other regions. In general, the bridge was evaluated positively in terms of revitalizing the island economy and improving island awareness but concerns were expressed about changes in morality, identity, and traditional culture resulting from the influx of Western culture from the metropolitan city, Surabaya.

We interviewed islanders about the effects that came about after the construction of the bridge and would like to discuss one case (Fig. 5.4). Before the construction of the bridge, one resident reported receiving IDR30,000 a day (about 2.5 USD) as a porter at a pier, but after construction he said he opened a store and earned IDR200,000 (about 18 USD) a day during the peak season. Now, many islanders have opened illegal shops around the bridge to sell products to tourists.

The surroundings will be tidied up soon, but the construction of the bridge has given the Madura islanders a huge profitable opportunity for the time being. Indonesia, which has enormous resources such as oil and timber has a very serious problem of wealth imbalance due to the corruption and illegal activities of public officials. With the construction of the bridge, a tremendous new town and resort are under construction on the beaches of Madura and Surabaya. Wherever there is development, there is shadow development, and the natives will always be under the shadow. Attention should be paid to how the future of Pulau Madura connected to the mainland of Indonesia will change and how the strong and unique living culture and the use of natural resources of the Madurese, who even the Javanese fear, will transform.

In the Bangkalan regency, before the establishment of the Suramadu Bridge, agriculture was the main sector in the local economy. After the Suramadu development, however, the economic activity has increased. Economic activities now include agriculture, trading, hotels, restaurants, transportation, communication, finance, rentals, and other business related to services. Studies by Yanti (2013) and Effendi and Hendarto (2014) found similar evidence. The fact that few young people are showing interest in agriculture now may lead to potential problems with the agricultural sectors in Bangkalan in the future.

Compared to the other regencies on Madura Island (Sampang, Pamekasan, Sumenep), Bangkalan is the nearest area to the metropolitan city in East Java (Surabaya). With the fact that the number of farmers and available lands are high, programs related to the strengthening of agriculture with a modern approach is needed. The potential programs for agricultural revitalization include empowering stallholders in modern agricultural practices, funding for agricultural infrastructure to support crop productivity, and training in good agricultural practices and entrepreneurship.

The rapid development of service industries in Sampang, however, is viewed as a potential factor for decreasing local social value. The local communities in Madura are practicing Muslims with a strict implementation of Islamic teachings. Some respondents highlight that rapid infrastructure development in Madura as a consequence of the Suramadu Bridge construction could potentially contribute to behavioral changes of local people in Pulau Madura. Rapid modernization will potentially change local values. Tourism development is a double-edged sword with negative and positive sides which are crucial for the future of the local community in Pulau Madura. The Muslim informal leader, the "*ulama*," notes that tourism development will bring a negative impact. This was supported by an elementary teacher who stated that the rapid development caused by the Suramadu Bridge could potentially affect the behavior of the younger generation.

5.4 Conclusion

Islandness is the unique identity of an island. This identity not only reflects the characteristics of the island's own physical and ecological resources, but is also determined by the socio-economic activities of the islanders who use these resources. When the shape of the island changes, the islandness also changes. Islandness is passed down and develops through the generations. The evolution of such islandness takes a long time with environmental changes and socio-economic changes. Recently, as construction of bridges connecting islands to the mainland has increased, the issue of island identity is increasing. Bridge construction has many convenient advantages for long-isolated islanders but they go through a process that makes them harmonize with the culture of the mainland (or cities). The connection between the island and the mainland by a bridge accelerates the change in the identity of the island. The islanders experience many cultural changes that they have never experienced in their lifetime and, over time, the traditional community and society change and disintegrate.

The construction of the Suramadu Bridge in Indonesia has improved the accessibility between Pulau Madura and Surabaya City, which is a positive economic effect. However, westernized Surabaya youth culture has influenced the traditional society of Pulau Madura. Some residents appreciate this cultural change in the island society but this is a concern from the perspective of other Pulau Madura residents, who have maintained their own culture for a long time. In Japan's Seto Inland Sea and Korea's archipelago, discussions on the changes in island society after the construction of the bridges are ongoing. We hope that Pulau Madura will take on board these examples to preserve its unique islandness to hand down to future generations.

Acknowledgment This research was carried out with the support of the Humanities Korea Plus (HK+) project of the National Research Foundation of Korea (2020S1A6A3A01109908).

References

Conkling P (2007) On islanders and islandness. Geogr Rev 97(2):191–201

Effendi M, Hendarto RM (2014) Dampak Pembangunan Jembatan Suramadu Terhadap Perekonomian Pulau Madura (Studi Kasus Kabupaten Bangkalan). Diponegoro J Econ 3(1):185–197

Hasan F, Darwanto DH (2013) Prospek dan tantangan usahatani tembakau madura. SEPA Jurnal Sosial Ekonomi Pertanian dan Agribisnis 10(1):63–70

Hong S-K (2010) Biocultural diversity and traditional ecological knowledge in island regions of Southwestern Korea. J Ecol Field Biol 34(2):137–147

Hong S-K (2011) Eco-cultural diversity in island and coastal landscapes: conservation and development. In: Hong S-K, Wu J, Kim J-E, Nakagoshi N (eds) Landscape ecology in Asian cultures. Springer, Tokyo, pp 11–28

Hong S-K (2015) Local activation using traditional knowledge and ecological resources of Korean islands. J Ecol Environ 38(2):263–269

Hatfield SC, Hong S-K (2019) Mermaids of South Korea: Haenyeo (women divers) traditional ecological knowledge, and climate change impacts. J Mar Island Cult 8(1):1–16

Hong S-K, Nakagoshi N (2017) Landscape ecology for sustainable society. Springer, Cham

Hong S-K, Pungetti G (2012) Marine and island cultures: a unique journey of discovery. J Mar Island Cult 1(1):1–2

Hong S-K, Bogaert J, Min Q (2014a) Biocultural landscapes: diversity, functions and values. Springer, Dordrecht

Hong S-K, Lee H-J, Kang B-R, Kim J-E, Lee K-A, Kim K-W, Jang D-H (2014b) Challenges and goals of the sustainable island: case study in UNESCO Shinan Dadohae Biosphere Reserve, Korea. In: Nakagoshi N, Mabuhay JA (eds) Designing low carbon societies in landscapes. Springer, Tokyo, pp 145–162

Hong S-K, Won Y-T, Lee G-A, Han E-S, Cho M-R, Park H-Y, Kim J-E, Chisholm Hatfield S (2018) Interdisciplinary convergence research design on island biocultural diversity—case study in Wando-gun (County) Island region, South Korea. J Mar Island Cult 7(1):12–37

Izzah L, Sulistiyono ST, Rochwulaningsih Y (2021) Commodifying culture in a frontier area: the utilization of Madurese culture for developing tourism in the eastern tip of Java Island, Indonesia. J Mar Island Cult 10(1). https://doi.org/10.21463/jmic.2021.10.1.072021

Kosim MK (2012) Kyai dan blater (elite lokal dalam masyarakat Madura). KARSA J Soc Islam Cult 12(2):149–160

Lee K-Y (2020) Folklore of the island, read with four keywords, vol 154. Minsokwon Archebooks, Seoul

Maffi L, Woodley E (2010) Biocultural diversity conservation—a global sourcebook. Earthscan, London

Suwarso S, Herwati A, Rochman F, Slamet S, Supriyadi B, Yasin B (2020) Varietas Hibrida Harapan Tembakau Madura. Jurnal Penelitian Tanaman Industri 8(1):27–31

Wekke IS, Ladiqi S, Bustami R (2019) Bugis and Madura migration in Nusantara: religiosity, harmony, and identity from eastern Indonesia. Ulul Albab 20(1):1

Whitten AJ, Whitten T, Soeriaatmadja RS, Soeriaatmadja RE, Afiff SA (1996) The ecology of Java & Bali, vol 2. Oxford University Press, Oxford

Yanti ATD (2013) Dampak Kebijakan Pembangunan Jembatan Suramadu Terhadap Sosial Ekonomi Masyarakat Dalam Pengembangan Wilayah Jembatan Suramadu (Studi Di Desa Sukolilo Barat Kecamatan Labang Kabupaten Bangkalan). Jurnal Administrasi Publik 1(2):147–154

6 Sustainable GeoHeritage Tourism: Bridging GeoHeritage and Culture Through the UNESCO Global Geopark Framework

Lee Jing, Sharina Abdul Halim, and Tanot Unjah

Abstract

The declaration of 2017 as the International Year of Sustainable Tourism for Development by the United Nations General Assembly heightens the significance and importance of tourism for the advancement of the universal 2030 Agenda for Sustainable Development. In the context of sustainable tourism, the framework of Global Geoparks would be well-placed in maximising the potential of geoheritage in fostering economic prosperity and social inclusiveness while at the same time, promoting peace and understanding through a mutual exchange of the richness, diversity and inherent value of nature and culture of the people. The interconnectedness between a geoheritage and the biocultural landscape in which it situates is manifested in the intertwinement of the day-to-day lives of the local communities and their interdependence with their land and the surrounding environment. This chapter illustrates the inextricable link between a geoheritage site and the local customs, practice, culture and way of life of a biocultural landscape, i.e., the Langkawi Geopark Community where the community is actively engaged in articulating their respective perspectives and dimensions of their interactions with their natural environment. The inclusive and participatory process, supported by the local authorities and all relevant stakeholders via an enabling governance framework, empowers the local communities where they are given an opportunity to exercise ownership over the decision-making process in charting the direction of a geoheritage conservation and the manner in which they could contribute towards the promotion of sustainable tourism. The bottom-up approach allows the co-generation of knowledge and more importantly, to instil

J. Lee · S. A. Halim (✉) · T. Unjah
Institute for Environment and Development (LESTARI), Universiti Kebangsaan Malaysia, Bangi, Selangor Darul Ehsan, Malaysia
e-mail: sharinahalim@ukm.edu.my; tanot-u@ukm.edu.my

S. A. Abdullah et al. (eds.), *Conserving Biocultural Landscapes in Malaysia and Indonesia for Sustainable Development*,
https://doi.org/10.1007/978-981-16-7243-9_6

in them a sense of pride that they are part of the Geopark community, which is instrumental in ensuring the viability and success of a Geopark as a geotourism attraction.

Keywords

Biocultural landscape · Cultural heritage · Custodian · Langkawi Global Geopark · Sustainable geotourism

6.1 Introduction

The 2030 Agenda for Sustainable Development adopted by the United Nations General Assembly (UNGA) on 25 September 2015 re-emphasised the importance of the sustainability of the planet in enabling social and economic development where it expressly declared that one of the ways to sustainably manage the planet's natural resources for sustainable development is sustainable tourism, which should be promoted.

In fact, the United Nations General Assembly had previously adopted Resolution 69/233 on 19 December 2014 that specifically addressed the 'promotion of sustainable tourism, including ecotourism, for poverty eradication and environment protection' that identified sustainable tourism as a 'cross-cutting activity that can contribute to the fight against poverty, the protection of the environment and the promotion of sustainable development', particularly for the promotion of 'rural development and better living conditions for sustainable rural populations' (UNGA 2015).

It is apt to link tourism with development given that the travel and tourism industry has an interest in the selling of 'the environment, both physical and human, as its product' (Murphy and Price 2005). There is solid evidence that supported the nexus between tourism and economic development, where it is increasingly recognised that the tourism industry contributes significantly towards the world economy. The World Travel and Tourism Council (WTTC)'s latest research, which was conducted in conjunction with Oxford Economics, revealed that the travel and tourism industry contributed to a total of 10.2% of the world's Gross Domestic Product (GDP) in 2016, which amounted to approximately US $7.6 trillion, and kept 292 million people in employment (WTTC n.d., unpublished).

The economic impact of travel and tourism to the world was broadly defined by the WTTC to include direct, indirect and induced impacts (WTTC 2017). According to WTTC (2017), the direct impacts of travel and tourism to the economy of a country could be seen through two perspectives, namely the impacts on commodities, which include accommodation, transportation, entertainment and attractions and their relative industries such as the accommodation services, food and beverage services, retail trade, transportation services, as well as cultural, sports and recreational services where the direct sources of funding came from domestic spending by residents, businesses, visitor exports and the spending of governments on travel and tourism.

In addition, the wider impacts of travel and tourism on the economy, which encompass indirect and induced contributions of travel and tourism towards the total contribution of the sector are calculated by WTTC in its 2017 Report. The 2017 Report elaborated that the increase in travel and tourism activities will indirectly lead to:

- Higher domestic purchases of goods and services by sectors directly connected to travel and tourism.
- An increase in investment spending such as the procurement of new aircrafts or the construction of new hotels.
- Higher collective expenditure of the government for the promotion of travel and tourism activities for the benefits of the community at large such as the tightening of security services especially in resort and attraction areas (WTTC 2017).

The direct and indirect economic contributions of the travel and tourism sector would induce the direct and indirect spending of employees involved in the sector on food and beverages, recreation, clothing, housing and household goods, which contributed towards the economic development of the area where travel and tourism activities are robust.

Although economic prosperity lies at the forefront of tourism, the importance of tourism, which is essentially an activity that encourages inter-cultural dialogue and exchange, to a certain extent, goes beyond economic development. Despite some scepticism raised on the role of tourism in enhancing peace (Var and Ap 1998), tourism has increasingly been seen as an instrument that could play a pivotal role in catalysing human interaction that advances 'understanding, tolerance and solidarity among all civilisations, peoples and culture' (UNGA 1999). Increased human interaction through tourism, where exchanges took place between peoples from different geographical and socio-cultural background on the site of a travel destination or a place of attraction, could lead to the cultivation of spiritual and cultural understandings and respects among and between peoples (UNWTO 2011), which is fundamental to the fostering of a culture of peace.

Having shown the important nexus between tourism and economic development, and substantiating the perception that sustainable tourism would be a viable vehicle towards achieving the Sustainable Development Goals (SDGs), with particular attention drawn to the contribution of tourism in spurring economic growth, it is timely to discuss the role of international platforms, in the context of the present chapter, the United Nations Educational, Scientific, and Cultural Organisation (UNESCO) Global Geopark Programme, in spearheading sustainable tourism through geoheritage tourism that contributes towards sustainable development. The chapter will first introduce the model of geoheritage tourism promoted by UNESCO Global Geopark as an innovative alternative mode of income generation and economic growth, demonstrated through the case study of Langkawi Global Geopark as an apt example of a biocultural landscape. Highlighting the importance of community participation in ensuring the success of a geotourism destination, the chapter seeks to articulate the inextricable link between Langkawi Global Geopark as a geoheritage tourism destination, and the local customs, practice, culture and way of life of the Langkawi Geopark Community through the construction of a community-oriented, and community-led narrative of cultural heritage particular to the Langkawi Island—a UNESCO Global Geopark.

6.2 Langkawi Global Geopark

The Island of Langkawi (Fig. 6.1), once known as the 'Isle of Legends' (Abdul Razak 2010), was under-developed for many years until it was accorded a duty-free status in 1987 (Hashim et al. 2011). Since then, the island underwent rapid development and in a short span of 20 years, the sleepy fishermen village transformed into a busy tourist town with hotels, infrastructures and facilities built across the island in support of the burgeoning travel and tourism industry. Apart from shopping, tourists from the country and beyond flocked to the Langkawi Island for various events such as the Royal Langkawi International Regatta, the KFC Langkawi International Regatta Perdana, the international cycling race, *Le Tour de Langkawi*, the Langkawi International Maritime and Aerospace Exhibition (LIMA), the Tuba Trail Run, the Langkawi International Ironman Triathlon and many more (Naturally Langkawi 2017).

The Langkawi Tourism Blueprint, launched by then Malaysia's Prime Minister in 2012 had outlined a short-term action plan that aimed to boost the island's travel and tourism industry (Mohd Yusof et al. 2014) by highlighting the Island's natural and cultural attractions, up from the ridges of Mt. Machinchang where tourists could have an expansive, panoramic view, from the cable car, of the forest canopy of tropical virgin forests, down to the coral reefs found at the Payar Island Marine Park, nicely wrapped up with local myth, legend and folklores, most notably—of the poignant blessing of Princess Mambang Sari who laid the body of her baby who died after 7 days after birth to rest at the Lake of the Pregnant Maiden; the tragic death of

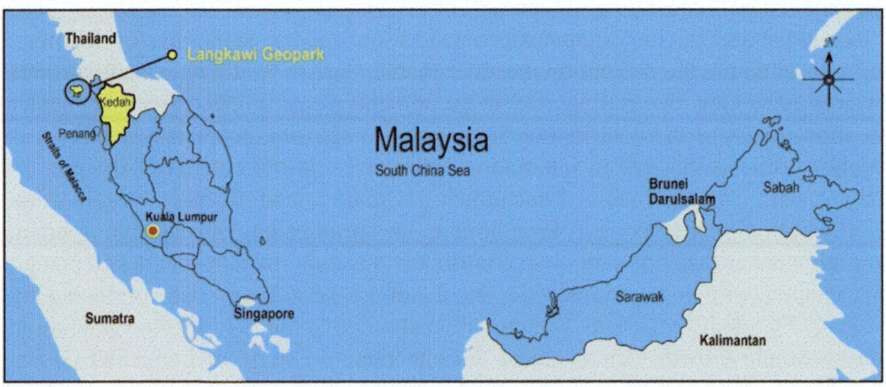

Fig. 6.1 The Island of Langkawi—the 'Isle of Legends' is located at the northern part of Malaysian peninsular

the beautiful Mahsuri, killed by her own ceremonial sword; and the ferocious fight of clashing pots and pans between the two feuding giants, Mat Chinchang and Mat Raya (Abdul Aziz and Ong 2011). Equipped with the mystical allure of beautiful maiden, princess, giants and all things magical, the unique geological landscape that is of outstanding universal value, beautiful beaches and exceedingly rich biological diversity endemic to the Island (Mohamad and Ahmad 2010), Langkawi Island has all it takes to be among the top ten best island and eco-tourism destinations.

Extensive investment was made in commercialising the travel and tourism industry of the Langkawi Island in order to promote the Island as a premier tourism destination in the region (Ong and Halim 2011). However, this raises concern over the mounting pressure exerted on the natural resources of the Island due to a large-scale construction of infrastructures and the spike in population brought forth by the flood of tourists visiting the Island, where mass tourism leveraging on the existing natural and cultural advantages without a careful study of the Island's carrying capacity may jeopardise the integrity of vulnerable bio-geo heritages and the conservation of the Island's natural resources as a whole (Ali and Unjah 2011). The rapid economic growth brought about by the avalanche of tourists coming to the Island and the subsequent mushrooming of facilities and infrastructures that usher in an even greater number of tourists to the Island do not spare the Island of the negative side effects resulted from the unchecked development—the damage to the environment due to pollution and the lack of an effective waste management system that is capable of catering to the escalating number of tourists—all of which compromise the ability of the Island's natural capacity to cope and accommodate such interferences (Hashim et al. 2011).

Moreover, the development of the Island, intensely focused on enabling and facilitating the travel and tourism industry, has so far concentrated only in areas that attract most tourists—the iconic places of attraction that are most visited, among others, the Kuah Town famous for its duty free shops and cheap souvenirs; the Mahsuri Tomb in memory of the legendary Mahsuri that put a curse on the Island for seven generations; the beautiful Chenang Beach along the coast and the Lake of the Pregnant Maiden (Mohd Ayob et al. 2013), while some areas of the Island remained under-developed where the majority of the population at these areas are still relatively poor, being marginalised from the rapid development unlike the other more popular areas (Hashim et al. 2011), whereby the 'trickle-down effect of growth that benefits and reaches poor and vulnerable groups takes time and effort due to the degree of accessibility of groups to resources, social and physical infrastructures and inadequate achievement in education and technical skills' (Abdul Halim et al. 2011).

Bearing in mind the importance of an inclusive economic development, and the urgent need to safeguard and conserve the integrity of the island's ecosystem in line with a global aspiration towards achieving sustainable development, the UNESCO Global Geopark Programme presented an exceptionally suitable paradigm in charting a sustainable development of the Island, especially when the Island is endowed with:

- a complete Palaeozoic geological succession incorporating the oldest rocks and fossils in the region, best preserved sedimentary structures and fossils, best sedimentological and palaeontological evidences for affiliation with Gondwana land; and
- the most beautiful island karst landscape in the region featuring unique hills, ridges, islands and pinnacles, beautiful caves, tunnels and arches and the magnificent rare mangrove associated with limestone bedrock (Leman et al. 2007).

The Global Geopark approach, defined to encompass a territory that comprises 'a certain number of geological heritage sites' of any scale, or 'a mosaic of geological entities of special scientific importance, rarity or beauty, representative of and area and its geological history, events or processes', which is not 'solely be of geological significance but also of ecological, archaeological, historical or cultural value', and more importantly, a delimited territory large enough to 'serve local economic development' gave greater emphasis on the interactions between socio-economic and cultural development and the conservation of the natural environment (Abdul Aziz et al. 2011; Leman et al. 2007). Abdul Halim et al. (2011) cited Kilim Geoforest park as an example of a pilot project that adopts the Geopark paradigm with the aim of improving local livelihoods, especially through the provision of innovative job opportunities for the local people, for example, creating opportunities for fishermen to participate in the tourism industry by becoming boatmen and nature guides, while at the same time, continue to fish in a manner conducive to the sustainability of the resource. By bringing more tourists to the existing natural assets located in the under-developed areas of the Langkawi Global Geopark such as the Wang Buluh and Wang Lebah caves on Tuba Island through the various geotourism initiatives under the Geopark paradigm, these under-developed areas could be made more accessible and in return, be able to enjoy the many benefits that a Geopark brings (Abdul Halim et al. 2011).

The underpinning philosophy of the Geopark initiative that advocated for the protection and sustainable development of geological heritage and geodiversity added a new dimension to existing paradigms of conservation where the potential for interactions between socio-economic and cultural development is highlighted in parallel with the conservation of the natural environment (Hashim et al. 2011). The balanced approach also resonates well with the aim of both the Federal and Kedah State governments to leverage on the potential of Langkawi Island as a world-class tourist destination in spearheading economic growth in the region through tourism (Ong and Halim 2011). The three-pronged approach of conservation, education and geotourism, laid down in the six criteria incorporated in the framework of the UNESCO Global Geopark Programme—size and setting; management and local involvement; economic development; education; protection and conservation; and last but not least, a global network, which should be satisfied in order to be granted the recognition as a geopark, ensures a sustainable development of the Langkawi Island (Abdul Aziz et al. 2011), while at the same time plays to the strength of the Island that is already renown as one of the most sought after tourism destination in the region.

When the whole of Langkawi's 99 islands amounting to a total land area of 478 km^2 was granted the status of Global Geopark Network (GGN), and endorsed by UNESCO in June 2007, the first in Malaysia and Southeast Asia subsequent to the concerted effort of the Malaysian Geological Heritage Group who had worked tirelessly in unveiling the geological secrets of Langkawi and the active advocacy in promotion of the unique and internationally significant geological features found on the Island (Ali and Unjah 2011), with the support of the Board of the Langkawi Development Authority (LADA) and the Kedah State government (Leman et al. 2007), it was found evident that the establishment of an enabling governance architecture that brings all relevant stakeholders together would be instrumental to the effective management of the Geopark was recognised (Abdul Aziz et al. 2011). In fact, the existence of a management plan 'designed for sustainable socio-economic development and demonstrate methods for conserving and enhancing geological heritage with broadening environmental issues' that is proposed jointly by 'public authorities, local communities and private interests acting together' is instrumental to the application for and recognition of a Global Geopark.

Taking into consideration that geoparks and land are one and the same, it serves both theoretical and practical purposes that the existing legal framework governing land use planning would be the most appropriate starting point in designing a governance architecture that is capable of capturing all the aspects and components entailed in the effective management of a geopark (Aziz et al. 2011). According to Aziz et al. (2011), in order to satisfy the requisites stipulated under the Operational Guidelines for UNESCO Global Geoparks, the existing regulatory mechanisms for land use planning provided under the relevant national legislations could be interpreted in a manner that enables the effective management of the geopark that seeks to balance conservation and development through the four 'building blocks' of:

- Boundary demarcation (the relevant legislations being the National Land Code 1965 and the Town and Country Planning Act 1976).
- Designation of authoritative body/institution (apart from the two legislations mentioned above, the Ministerial Functions Act 1969).
- Measures that drive sustainable tourism and sustainable economic development (the Lembaga Pembangunan Langkawi Act 1990 and the Town and Country Planning Act 1976).
- Regulatory mechanisms that effect the conservation of the geological, biological and cultural heritage and area (the Town and Country Planning Act 1976, the National Land Code 1965, if necessary, the Land Acquisition Act 1960 (revised 1992), the National Forestry Act 1984, as well as the National Heritage Act 2005).

By drawing together the salient aspects crucial to the effective management of a geopark, land use planning could serve as a 'powerful means to seek to recognise boundaries, set out "controls" to facilitate conservation and "guide" development, advocate sustainable tourism and economic development as well as ensure heritage is protected, conserved and served as means to educate and inculcate a sense of place' (Aziz et al. 2011), and ultimately, achieve the overarching aim of sustainable development by leveraging on the paradigm of geotourism advocated under the UNESCO Global Geopark Programme. More importantly, the management plan should be structured on a governance platform that is facilitative of a decision-making process participated effectively by local communities, private interests, all relevant stakeholders and interested parties, which demonstrates 'respect, encouragement and protection of local cultural values' for the proper management of a UNESCO Global Geopark, and in the present case, the Langkawi Global Geopark (Ong et al. 2010).

6.3 Methodology: Articulating the Nature-Culture Linkage from the Bottom-Up

At this juncture, it is pertinent to reiterate the demand imposes on the governance architecture established for the management of a UNESCO Global Geopark. The geopark paradigm requires the incorporation of a bottom-up approach (Operational Guidelines for UNESCO Global Geoparks, Paragraph 1) that sufficiently reflects the spirit of inclusion, participation and equity, whereby processes and procedures that encourage participation and engagement of local authorities, all relevant stakeholders, interested parties and the society at large in relation to the management of the geopark should be established. The Operational Guidelines for UNESCO Global Geoparks, attached to the UNESCO Statutes of the International Geoscience and Geoparks Programme, made explicit that in order to be accorded the status of a UNESCO Global Geopark, the geopark must be 'managed with a holistic concept of protection, education, research and sustainable development' by

> a management body having legal existence recognised under national legislation . . . that is appropriately equipped to adequately address the area of the UNESCO Global Geopark in its entirety (Paragraph 3 on the Criteria for UNESCO Global Geoparks).

The Operational Guidelines, under the same Paragraph, further provide that the management of the Geopark should be sufficiently represented by all relevant local and regional actors and authorities and allow for the active involvement of 'local communities and indigenous peoples as key stakeholders in the Geopark' through a co-management plan that incorporates local and indigenous knowledge alongside science, which was drafted and implemented to serve 'the social and economic needs of local populations, protects the landscape in which they live and conserves their cultural identity'.

The requirement to ascertain the linkages between socio-economic and cultural development with the conservation of natural heritages through geotourism is articulated in Category 1.3 of the Self-Evaluation Form for Aspiring UNESCO Global Geoparks on Natural and Cultural Heritage where the applicant states are asked to self-evaluate their efforts to promote the links between geological heritage and other natural and cultural sites within the Geopark, including public awareness, capacity building and education programmes that communicate and disseminate geo-scientific knowledge and environmental needs and concepts (Abdul Aziz et al. 2011). The Self-Evaluation Form requests for:

- Details for the interpretation, communication and education programmes that were implemented in order to achieve the objective of promoting and maintaining the relevant natural and cultural heritage (Sub-Category I.3 of the Self-Evaluation Form for Aspiring UNESCO Global Geoparks).
- Details regarding the management structure organised for the effective management of Global Geoparks, especially the existence of initiatives or working groups that discuss the promotion of natural and cultural heritage (Category II of the Self-Evaluation Form for Aspiring UNESCO Global Geoparks).
- Details of research, information, education and other scientific activities on Earth Science within the Geopark territory, and in general the operationalisation of education programmes concerning the Geopark in the area (Category III of the Self-Evaluation Form for Aspiring UNESCO Global Geoparks).

Keeping in view that the geopark approach aspires towards a knowledge-based sustainable geotourism that is inclusive and people-oriented (Leman et al. 2007), it is incumbent to review the self-evaluation criteria stipulated under the Self-Evaluation Form for Aspiring UNESCO Global Geoparks, particularly Categories I (I.3), II and III from a bottom up perspective. Instead of a scientists-led, top-down interpretation, communication and education of matters relating to a UNESCO Global Geopark that were translated into a process of 'informing and educating' the people about the salient concepts and knowledge of geo-science and the environment (which could be patronising at times), the scientists assume a facilitative role in soliciting the perspectives of the Geopark Community—the Geoparkians as to how they could contribute towards the interpretation of the nature-cultural linkage, and how they could play a role in communicating and educating the community (themselves) and others, of the local knowledge on the Geopark (Hashim and Abdul Aziz 2013). This aspect of community-led, bottom-up social learning, alongside the sciences, contributes towards the strengthening of the nature-culture linkage characteristics of the Geopark paradigm.

This process of community-oriented engagement allows the bilateral flow of information and knowledge, and at the same time, empowers the community in reinstating their identity and ownership vis-à-vis the geopark. Connecting a place with the identity of the society through an inclusive and participatory development process—from planning, implementing, to monitoring—is instrumental in inculcating pride and a sense of belonging within the community to the area. This feeling of ownership, togetherness and belonging ultimately leads to a wider acceptance of the UNESCO Global Geopark Programme by the people as a means to promote sustainable geotourism for sustainable development (Farsani et al. 2012).

Armed with this objective to further define the linkage between nature and culture within the framework of the UNESCO Global Geopark, a research team from the

National University of Malaysia set out to identify the key custodians of the cultural heritage of Langkawi Global Geopark and at the same time, map out the cultural heritage of the Island. The research endeavours to understand how the Geopark Community defines 'cultural heritage', in what way and how they identify themselves within the context of Langkawi as a UNESCO Global Geopark. The research built upon the ongoing effort to ascertain, understand and document both the natural and cultural heritage of Langkawi that has started formally since 2003 by the team of researchers at the Langkawi Research Centre. The underpinning philosophy in the charting of the research is, 'Making the Past Present for the Future' (Hashim et al. 2013) with its central focus on the peoples and their heritage, and the value common to, and shared among, the Langkawi Islanders (Hashim and Abdul Aziz 2013). Their work had made important contributions towards fulfilling the criteria and standards imposed in order to be accorded the status of Global Geopark, and the drafting of the Langkawi Geopark Management Plans.

The present research improves on the previous research by directing the scope of research to the unravelling of the implicit dimension of the people's cultural heritage, where the research seeks to profile the types of cultural heritage, both tangible and intangible, in the three *mukims* (sub-districts) of the Island, namely Padang Mat Sirat, Bohor and Ulu Melaka, and to identify the relevant custodians of these cultural heritage. The research had identified several categories of tangible and intangible cultural heritage found in Langkawi Island. The tangible cultural heritage ascertained is preliminary placed under the themes of traditional village (*kampung tradisi*), historical sites, house of prayers (*rumah beribadat*), traditional craft and livelihood and craft, whereas the intangible cultural heritage found in the Island of Langkawi could be placed under the themes of food, traditional medicine, custom and culture, traditional games and the art of self-defence, performing arts, fine arts, and languages and writing.

This chapter presents a narrative of cultural heritage constructed from the outcomes and findings obtained from the interviews and focused group discussions with the artisan, Mr. Dun bin Chin, the custodian of one of the most prominent local cultural heritage in the Island of Langkawi—the building of Malay traditional house, the *rumah Melayu*. The Malay traditional house was chosen because it embodies both the tangible aspect of a cultural heritage—the house itself, and the intangible dimension of the cultural heritage, which are the skills, knowledge and craftsmanship involved in the building of the house. More importantly, since house is so instrumental to a society, the ways and manner in which a house is constructed will reveal not only the preference of people at that time, but also the customs, practices, culture and way of life of the people who reside in a house that could have been built by them and their neighbours who lived nearby.[1] The narrative seeks to illustrate

[1]The symbolic representation of what a house means to a culture was eloquently described in the questions that Janet Carsten (1997, p. 33) posed in her book, *The Heat of the Hearth. The Process of Kinship in a Malay Fishing Community*.

"What is it that gives the house in Langkawi its significance? What makes a house a house? Is it its spatial layout, its physical structure as a building, the rituals which are enacted in it and which are

how a bottom-up process provides a different perspective in which the criteria stated in the evaluation document of geopark, especially those that were specifically raised above, could be construed in a manner that is more conducive to the identification and the promotion of interlinkages between natural and cultural heritage, where cultural heritage is not viewed as a dimension complementary to the natural heritage, but instead, be treated as the soul that makes the place, and gives life, identity, characteristics to a place, which makes the place unique.

6.4 A Narrative of *Pak Don*: The Custodian of the Cultural Heritage of Traditional House Building

Mr. Dun bin Chin, or *Pak Don*, was born in Langkawi in 1946. He is a master of all trades, well-known in the community as the wood artisan who builds Malay traditional houses (the *rumah Melayu*). He was commissioned to build a *rumah Melayu* by the cousin of the Sultan of the State of Kedah in 2011, which took him a year to complete the commission. In addition, *Pak Don* is adept at all sorts of woodcrafts where he crafts the sheaths (*sarung*) that cover traditional Malay knifes (the *golok*) and makes traditional toys such as the spinning top (*gasing*). Apart from woodcrafts, *Pak Don* is also renowned for the production of *nira*, the palm juice of a coconut tree that is well-known for properties beneficial to health, which could be also be processed into various types of sweetmeat, including palm syrup or honey, and palm sugar (*gula Melaka*).

In realising the inextricable link between tangible and intangible cultural heritage, the construction of the traditional Malay house would be an excellent starting point for the study of what cultural heritage means from a community's perspective, and how the promotion and maintenance of the link between natural and cultural heritage could be achieved. Reiterating the centrality of a house to the Malay community, and the symbolic representation of a traditional Malay house of traditional Malay culture and lifestyle, the research team decides to document the whole process of building a traditional Malay house from scratch, and to ascertain the intangible dimension of the cultural heritage embedded in the construction of the traditional Malay house in order to study the relations between the tangible and intangible dimension of a cultural heritage. This exercise is important to gain a deeper understanding of the different facets of cultural heritage, and whether a cultural heritage would cease to be a 'heritage' if the values, beliefs, worldview, way of life, in short, the 'culture' that were embedded in the heritage, are lost.

For these reasons, the team of researchers placed a request to *Pak Don* to construct a 'miniature' *rumah Melayu* to learn about the philosophy behind the architecture and spatial arrangement and design of the house in order to fully understand the concepts adopted, and the processes and skills entailed in the

part of the process of building? Or the social significance of the house an aspect of the quality and types of relations of the people who live within it and of the activities which they engage in there?"

Fig. 6.2 The 'miniature' of a traditional Malay house, with *Pak Don*, its creator, looking at it fondly (Photo by SA Halim)

construction of a *rumah Melayu*. To everybody's amazement, which left us all in awe of the mastery of the craft and workmanship, *Pak Don* built the traditional Malay house with the blue plan all inside his head, including how much materials are needed. He took his time in designing how the house could be constructed, the size, outlets for ventilation, the overall structure and the building processes, in particular when the order of the construction of the miniature house would have to be reversed due to the reduction in scale. Instead of building the *rumah Melayu* from the outside in, he would have to build it inside out as the house would be too tiny to be worked on after the external structure is built. Apart from the main structure, *Pak Don* completed the accessories to the house as well, where he designed and made the mould for the cement stilts supporting the *rumah Melayu* and carved and engraved the fences, the windows and other various parts of the *rumah Melayu* that is traditionally ornamented. A picture of the *rumah Melayu* is shown below in Fig. 6.2.

It took *Pak Don* a bit more than half a year to complete the house. After the miniature house was built, *Pak Don* explained the various dimensions regarding each structure and spatial arrangement of the house, and how he constructed it. Since the miniature will be used as a teaching material, apart from being showcased as an item on exhibition for everybody who came to the Langkawi Research Centre, *Pak Don* incorporated the evolution in the trends of how *rumah Melayu* is built over the years into the design and construction of the *rumah Melayu*. He showed the differences between the first generation and second generation of *rumah Melayu*, and the theory of 'expansion' in the traditional way of building the *rumah Melayu*.

First to be built would be the mother house (*rumah ibu*) with a raised platform (*serambi*) slightly lower than the living room extended out from the mother house, a living room, and a bedroom, followed by the kitchen (*rumah dapur*) which is attached to the mother house. More 'houses' ('*rumah*') will be built to accommodate more needs or more people in the house, without changing the structure of the main house.

Together with his wife Azizah, lovingly known as *Che Tipah*, they laboured over every detail of the miniature house, from whether the roof should be made from wood or the traditional roofing material—the *nipah* tree leaves; the drainage system over the kitchen top made from bamboo (*para buang*); the airy loft over the kitchen for keeping freshly made Malays cakes (*kuih-muih*), known as '*para kuih*', which sometimes, serves as a hiding place for the unmarried maiden of the house to take a sneak look at the prospective beau who came to visit her parents to ask for her hands of marriage; a little bed (known as '*getar*') for the newly weds' room (*bilik pengantin*) at the back of the house; down to the tiniest interior features of tiny pink curtains hanging up on the little window and plaited pandanus leaves mats (*tikar*) covering the floor. They even managed to find an old pot ('*belanga*') completed with a ladle that are traditionally placed at the side of the staircase leading up to the house for the family or guests to wash their feet before they go up the stairs.

Pak Don and *Che Tipah* explained the purposes for each room and how the rooms are put to use; they elaborated the functionality of each feature of the interior design of the *rumah Melayu* and spiced up their explanations with tales and anecdotes that came from their experience. By learning and gaining more understanding about the structure and features of the *rumah Melayu*, the more we learn about the day-to-day lives of the people of Langkawi once upon a time, the values they hold dear, the social matrix at that time, the evolution of needs and preference and many more facets of human lives that we had chosen to forget or discard when we embrace development and modernisation.

Pak Don informed the team of researchers that, after determining how much wood and other raw materials sourced from the nearby Mountain Raya that he would need for the *rumah Melayu*, he would wait until he has enough demand for wood and other forest produce that justify the cutting down a tree or other types of vegetation, such as bamboo in the mountain. Unless and until he secured sufficient demand, for example, wood, he would not cut down a tree in order not to waste any part of the tree. 'Take only what is needed, and waste nothing'—sustainability at its core—is the underlying motto of *Pak Don* in executing his creation. Each timber and plank is put to good use. The design of the *rumah Melayu* is simple and yet functional, and at the same time, aesthetically pleasing, with intricate carvings and engravings serving as ornaments for the house.

The *rumah Melayu* is more than a tangible cultural heritage object that we could express our awe, or for us to marvel at the level of mastery and skills demonstrated in its construction. It is a physical manifestation of identity, custom, culture and way of life adopted and practised by the people who lived in the *rumah Melayu.* The intangible dimension of cultural heritage, in particular, the respect they had towards nature, and the cautious manner in which they treat their biological and geological heritage, the knowledge they had about their surroundings and the skills they had acquired across the generations, are invaluable and priceless, and should be preserved and pass on to the future generations at all cost. As the research unfolds, it slowly unveils the local community's perspectives of the interactions with their natural environment, reflected in the local customs, practice, culture and way of life, which indirectly illustrates the inextricable link between a geoheritage site and the Langkawi Geopark Community.

The active communication of the geopark paradigm, by the community themselves and facilitated by the researchers under the research project rekindles the importance of the protection and conservation of natural and cultural heritage for the promotion of sustainable tourism that could elevate the socio-economic well-being of the Langkawi community. The various mechanisms deployed under the umbrella of UNESCO Global Geopark to uphold, strengthen and promote the principle of the protection and conservation, education and the sustainable development of the natural and cultural heritage of man, which forms the bedrock of the spirit of 'kawi' (*semangat kawi*)—a root word of 'Langkawi', a term coined to represent the 'spirit of the eagle' (*lang* for *helang*, another root word for Langkawi) that is of a reddish brown colour ('*kawi*')—shared among the local community and the stakeholders who are involved in the development of Langkawi (Ong et al. 2010).[2]

6.5 Demonstrating the Natural-Cultural Linkage Through the Narrative of *Pak Don*, the Cultural Heritage Custodian of Langkawi Global Geopark

Referring to Section 3 of Category I (1.3) on Natural and Cultural Heritage of the self-evaluation form for Global Geoparks that stipulates for the criterion of promoting and maintaining natural and cultural heritage, Section 3.1 requested for details that demonstrates the promotion of the links between geological heritage sites and

[2]The excerpt of the original writing of the authors on the '*kawi*' spirit (*semangat kawi*) is:

"*'Kawi', salah satu kata asas dalam kata penuh 'Langkawi' dikatakan merujuk kepada sejenis batu berwarna 'merah' yang terdapat di Langkawi. 'Lang,' kata asas kedua, merujuk kepada helang, yang banyak terdapat di Langkawi. Helang melambangkan semangat membuat sesuatu berdasarkan tujuan yang jelas dan tepat pada sasaran. Justeru, tajuk Makala hini, 'menjejak semangat 'kawi': Langkawi sebagai Geopark,' cuba menjelajahi semangat helang yang juga berwarna seperti kawi (merah-kecoklatan) dalam kalangan komuniti tempatan dan golongan yang terlibat dalam pembangunan Langkawi, agar segala sumber geo, bio dan budaya di Langkawi dilindungi dan dipulihara.*"

(Translation: '*Kawi*' is a basic word in the full term of 'Langkawi' that was said to have referred to a type of rock that is of the colour 'red' found in Langkawi. '*Lang*' is the second basic word, which referred to eagle that could be found aplenty in Langkawi. Eagles represent the passion to do something that is based on a clear purpose and right on target. Hence, the title of the paper, 'tracing the spirit of '*kawi*': Langkawi as a Geopark,' attempts to revisits the spirit of the eagle that is also of a reddish-brown colour (the colour of '*kawi*') amongst the local community and the group of people who are involved in the development of Langkawi, so that all the geo, bio and cultural resources in Langkawi could be protected and conserved).

Table 6.1 Sch. 1 Category III (geology and landscape) 1.3 Natural and cultural heritage of the self-evaluation form

I. Geology and landscape	
1.3 Natural and cultural heritage	
1	**Natural rank**
1.1	International designation in part of the Geopark territory (except World Heritage Sites and Biosphere Reserves)
1.2	National designation in part of the Geopark territory
1.3	Regional designation in part of the Geopark territory
1.4	Local designation in part of the Geopark territory
2	**Cultural rank**
2.1	International designation in part of the Geopark territory
2.3	Regional designation in part of the Geopark territory
2.4	Local designation in part of the Geopark territory
3	**Promotion and maintenance of natural and cultural heritage**
3.1	Promotion of the links between Geological Heritage sites and the existing Natural and Cultural sites within the Geopark
3.2	Interpretation
3.3	Communication
3.4	Education programmes

the existing natural and cultural sites within the geopark. Category I (1.3) on Natural and Cultural Heritage of the self-evaluation form for Global Geoparks is reproduced in Sch. 1 (Table 6.1).

Prior to the satisfaction of this criterion, it is incumbent to ascertain and identify the existing natural and cultural sites that are connected to a geological heritage site. Although the classification of what constitutes a geological heritage site, a natural site and a cultural site is somehow artificial for the biological, geological and cultural triad is indivisible, being situated in the environment that itself forms a unitary whole, the profiling and mapping of a custodian, based on the place where the custodian resides permanently, could provide us with an idea as to where a 'cultural site' for an intangible cultural heritage could be in order to meet the requirements stated in the self-evaluation form.

The research conducted by the team of researcher from The National University of Malaysia, or Universiti Kebangsaan Malaysia (UKM) as it is commonly known profiled and mapped where a custodian stayed and identify the area as a cultural site for the purpose of studying a link between the three types of heritages. Instead of a top-down, conventional science-led determination of where and how the linkage is formed, the research initiated a bottom-up approach where the custodians themselves were given the opportunity to paint a picture of how they, as the custodian of a cultural heritage, are connected to the geological and natural heritage surrounding and instrumental to the practice of a cultural heritage. The narrative of *Pak Don* demonstrated that an inextricable relationship between the building of a *rumah Melayu*, the natural materials used in the construction of the *rumah*, such as the wood, bamboo, pandanus leaves and *nipah* from the mountain. The manner and processes in which the *rumah* is built will have an impact on the sustainability of the resource where a culture of sustainability would be translated into a sustainable use of resource. As such, the link between geological heritage sites and the natural and cultural sites, especially an intangible cultural heritage, could be demonstrated through the performance of the activity that applies the said intangible cultural heritage, which is, in the present case, the knowledge and skills involved in the building of the *rumah Melayu*, and the art and craft of woodworking.

If interpretation under section 3.2 of Category I (1.3) on Natural and Cultural Heritage of the Self-Evaluation form (refer to Sch. 1) is interpreted to encompass the interpretation not just of the geological heritage, but also other natural and cultural heritage that are connected to the geological heritage, then, the role of interpretation need not be played by just the scientist and the authorities. The present case had sufficiently showed that the custodian would be in a better place to interpret the cultural heritage of the building of a Malay traditional house, especially when he is the person who understands most the significance of the heritage item or object, e.g., the Malay traditional house that encapsulates the way of life of the Malay community in Langkawi, and the values embedded in the house—from the no-waste policy in the construction of the house, to the importance placed on cleanliness evident from the use of a pot beneath the stairs to allow the guests and family members going to the house to wash their feet before entering the house.

As the custodian, *Pak Don* elaborated all aspects of the traditional Malay house, the life of the people who used to reside in time unfolds before the eyes of the listeners. We seemed to relive the lives of the ancestors, and in some way, strengthens our understanding about who we are, and the identity that characterises our worldview and beliefs. In this context, the satisfaction of section 3.2 Category I (1.3) on Natural and Cultural Heritage of the self-evaluation form (presented in Sch. 1) for Global Geoparks need not came just from the scientific experts. Instead, a suitable platform should be given to the custodian of a cultural heritage to interpret a cultural heritage for the purpose of promoting the link between geological heritage and natural and cultural heritage.

Likewise, communication as stated under section 3.3 Category I (1.3) on Natural and Cultural Heritage of the self-evaluation form (presented in Sch. 1) could be civilian-led as well. When the miniature *rumah Melayu* was built, words travelled. People from all over the country came to take a look at the miniature. The heightened attention drawn by the miniature house would be a perfect opportunity for the communication of the cultural heritage, and at the same time, introduces and exposes the visitors to the rich and outstanding geological and biological heritage of Langkawi Global Geopark. Instead of an authority or scientific expert-led process of communication, a civilian-led, bottom-up communication of a cultural heritage could be undertaken, and perhaps, expands the potential and breath in which the requirement of 'communication' under section 3.3 (Category I (1.3) on Natural and Cultural Heritage of the self-evaluation form for Global Geoparks) is satisfied.

If *Pak Don* could be invited to conduct lectures and workshops on Malay culture and woodcraft on a regular basis and be incorporated formally into the curriculum of primary, secondary and tertiary education systems, the requirement of establishing education programmes regarding natural and cultural heritage, and the link between them as stipulated under section 3.4 (Category I (1.3) on Natural and Cultural Heritage of the self-evaluation form for Global Geoparks as outlined in Sch. 1) could be achieved by enabling the participation of the custodian in the leading and design of educational programmes that do not only interpret a cultural heritage and communicate to others about it, but could also form part of the educational programmes organised for the promotion of the links between the three types of heritage.

A civilian-led, bottom-up approach could be applied in the satisfaction of the criteria laid down under Category III on Information and Environmental Education as outlined in the Self-Evaluation Form. The main sections of Category III are:

1. Research, information and education scientific activity in Earth sciences within the territory
2. Do you operate programmes of environmental education in your Geopark area?
3. What kind of educational materials exist?
4. What kind of published information is available in your Geopark area?
5. Geology provision for school groups
6. Education—Guides
7. What kind of information do you provide to educational groups to encourage them to visit your area?
8. Do you use the internet for school programmes? What kind of service do you provide?

In addition to the stipulations of research, information and education scientific activity in Earth sciences within the territory in section 1, the operation of programmes of environmental education in the geopark area in section 2 and geology provision for school groups in section 5, all of which are found in Category III of the Self-Evaluation form, perhaps the determination of whether a Global Geopark status should be accorded to an applicant country should perhaps, provide for an informal mode of information dissemination and environmental education that is initiated from the grassroot.

In furtherance of this proposal, recognition should be given to grassroots efforts that disseminate and share information and knowledge about the environment that are not limited to just geological sciences. A comprehensive evaluation of Category III is important in realising the aspiration of the Geopark framework to advocate social inclusion that brings all the relevant parties on board. In order to do so, the criteria stipulated under Category III, which focus on Information and Environmental Education should also acknowledge civilian-led initiatives, educational programmes and other research, information and educational activities on the three types of heritages and the linkage between them in promotion of a more inclusive and participatory process provided under the Geopark framework.

It is important that the management structure for the management of Global Geoparks as outlined in Category II of the Self-Evaluation Form specifically provides for mechanisms that support and enable a participatory, bottom-up approach. As stated in section 3.8 of Category II, the establishment of community links should be formalised and executed through the support of legal instruments that impose the obligation of the authorities to involve and engage the community actively in all decision-making matters regarding the management of a Global Geopark, in the present case, the Langkawi Global Geopark. The governance architecture presented in the second part of the chapter explains the four building blocks for the effective management of the Langkawi Global Geopark, especially the third building block of—'measures that drive sustainable tourism and sustainable economic development (the Lembaga Pembangunan Langkawi Act 1990 and the Town and Country Planning Act 1976)'—that provides for the establishment of an appropriate platform that facilitates and enables public participation. Instead of providing for 'measures that drive sustainable tourism and sustainable economic development', the building block should also stipulate for measures *developed jointly by the authorities and all relevant parties, including but not limited to local communities, relevant stakeholders and other interested parties* that drive sustainable tourism and sustainable economic development under the Geopark framework.

One of the unexpected outcome of the adoption of the UNESCO Global Geopark Programme by the State of Kedah for the effective management of Langkawi Global Geopark would be the catalytic role that the programme played in enabling the observation of a principle of international environmental law that imposes an obligation on the state to ensure the participation of the public in the decision-making process on any matters that has an environmental implication, be it direct or indirect (Lee 2012). A specific requirement that demands the taking into account of 'community links' in the devise of a management structure for the effective management of a geopark as stated under section 3.8 of Category II would compel the authorities and policy makers to perform their obligation under this specification in order to attain the status of a UNESCO Global Geopark, and by doing so, forces the country to satisfy its obligation under international environmental law.

The observation of the obligation to ensure public participation in environmental decision-making has a far-reaching effect. It promotes legitimacy, transparency and accountability in decision-making. More importantly, by enabling the participation of the public in the decision-making process, it confers upon them a sense of ownership and belonging to the process and the decision-made, which generates a wider acceptance of decisions made over the management of geopark and ultimately ensures the success of the management of a geopark. By drawing the example from the narrative on *Pak Don*, the custodian of cultural heritage in, among others, the building of traditional Malay house and other woodcraft from the Island of the reddish-brown eagle—the Langkawi Global Geopark, it could be demonstrated that a civilian-led, bottom-up approach is equally effective and capable in articulating the link between natural and cultural heritage, and subsequently, supports the promotion and maintenance of the link thereof.

This chapter endeavours to propose an alternative interpretation of the UNESCO Global Geopark framework embodied in the Operational Guidelines as manifested in the criteria enumerated under the self-evaluation form, which furthers the promotion of a linkage between natural and cultural heritage, and in addition, realises the potential of the geopark framework as an appropriate platform that aspires to:

- [create a geopark that was initiated by the] local communities/authorities with a strong commitment to developing and implementing a management plan that meets the community and economic needs of the local population while protecting the landscape in which they live;
- Involve public authorities, local communities, private interests and both research and educational bodies, in the running of the geopark and its regional economic and cultural development plan and activities. This cooperation shall stimulate discussion and encourage partnerships between the different groups having a vested interest in the area and motivate and mobilise local authorities and the local population and
- [Carry out] sustainable tourism and other economic activities within a geopark [successfully with the cooperation of] local communities. Tourism activities have to be specially conceived to match local conditions and the natural and cultural character of a territory and must fully respect the traditions of the local populace. Demonstrable respect, encouragement and protection of local cultural values is a crucial part of the sustainable development effort (Ong et al. 2010).

Following the proposal advocated for the articulation of the link between natural and cultural heritage spearheaded by the custodians and the civilians, which modifies the essentially top-down, authorities or scientific experts-led paradigm of the geopark approach, the chapter seeks to argue that a geopark could also be deemed to be successful not just by measuring the number of tourists who flocked to the island to look at the magnificent geological landscape found on the island, and by so doing, generates income for the local community; but also by measuring the number of tourists who flocked to the Island to observe the beautiful traditional Malay house (among other cultural and natural heritage objects), and at the same time, be exposed to the magnificent geological heritage that could be found on the Island, which similarly, generates income to the local communities and improves the socio-economic conditions of the local population.

6.6 Conclusion

The chapter presents the opportunity provided by the Global Geopark Programme in enabling a platform that allows the incorporation of the '*kawi*' spirit (*semangat kawi*), a shared commitment, principle and value of the Langkawi Geopark community towards the protection and conservation of the natural and cultural heritage of the Langkawi Global Geopark, while at the same time, benefits from improved socio-economic conditions brought forth by the geotourism industry. The various mechanisms stipulated under the geopark framework in promotion of a more robust geotourism industry empower the geopark community to exercise more ownership in the decision-making process in charting the direction of growth of the Langkawi Global Geopark. The potential of the Global Geopark Programme in bringing geoheritage and the society together in the same bandwagon towards sustainable development, where sustainable geotourism is touted as the innovative product that could improve socio-economic conditions of the geopark community, is illustrated through a purposive, civilian-led, bottom-up interpretation of the requirements stipulated under Category I, Sub-Category 1.3(3), and Category III of the Self Evaluation Document that specifically address the various aspects relevant to the promotion of the links between geological heritage sites and the existing natural and cultural sites within the Geopark, and the educational dimension in relation to the development of a knowledge-based geotourism industry, within the context of the Langkawi Global Geopark.

It is noteworthy that the partnership between science and society advocated under the Global Geopark Programme stimulates the bilateral flow of information and knowledge between the scientists and the community. The participatory process promotes social inclusiveness, which is crucial in facilitating a sense of ownership and belonging—that the local community, all relevant stakeholders and other interested parties—are part of, and belonged to the Langkawi Global Geopark. The feeling of 'we belong together' is instrumental in ensuring a wider acceptance of the geopark paradigm by the community that determines the viability and success of the Langkawi Global Geopark as a geotourism attraction. More importantly, the feeling of ownership in the development process through sustainable geotourism could be fostered in the process of public education, awareness raising and capacity building where the geopark framework provides a platform for social learning that instils in the community of the geopark (the Geoparkians) a sense of pride that they live in an area endowed with unsurpassable geological, ecological, archaeological, historical and cultural heritages of outstanding universal value, and ignites their desire to protect and conserve Langkawi Global Geopark's precious natural and cultural heritage.

Acknowledgement The authors would like to express their sincere gratitude to Latifah binti Md Yusoff and Siti Munirah binti Haji Abdul Hamid for their editorial assistance rendered in the preparation of this chapter. This chapter forms part of the outcome of a research that identifies and maps out the cultural heritage of the Island of Langkawi conducted under the Fundamental Research Grant Scheme awarded by the Ministry of Higher Education, Government of Malaysia (Research Grant No.: FRGS/1/2014/STWN09/UKM/02/1). Last but not least, the authors would like to extend their appreciation to the Langkawi Research Centre for the support provided in facilitating the research. Part of this paper has been presented at the *2017 International Geopark Conference, Taiwan: Communicating Geopark, How Local Involvement Makes a Better Tomorrow Possible*? organised by the Geopark Association Taiwan.

References

Abdul Aziz R, Ong PL (2011) Placemaking, place names, and local myths and legends. Plan Malays 1:145–158

Abdul Aziz R, Hashim HS, Komoo I (2011) Geopark for heritage conservation: a need for integrated planning and management. Plan Malays 1:25–38

Abdul Halim S, Komoo I, Salleh H, Omar M (2011) The geopark as a potential tool for alleviating community marginality. Shima Int J Res Island Cult 5(1):94–113

Abdul Razak N (2010) From fairy princess to a wronged maiden: assessing self-representation of langkawi as the 'Isle of Legends'. In: Proceedings of Regional Conference on Tourism Research (RCTR 2010): The State of the Art and Its Sustainability, pp 338–348

Ali CA, Unjah T (2011) The importance of geological heritage resources in land use planning: experience from Langkawi Geopark. Plan Malays 1:55–82

Aziz S, Hashim HS, Unjah T, Abdul Aziz R, Chan KLG (2011) Land use planning statutes for Langkawi Geopark conservation and development. Plan Malays 1:83–102

Carsten J (1997) The heat of the heart. The process of kinship in a Malay fishing community. Clarendon Press, Oxford

Farsani NT, Coelho C, Costa C, Carvalho CND (2012) Geoparks and geotourism: concepts, theories and paradigms. New approaches to sustainability for the 21st century. Brown Walker, Boca Raton

Hashim HS, Abdul Aziz R (2013) 2.0 Asas Pelan Tindakan, Pelan Pengurusan Langkawi Geopark (PPLG) 2012-2030. Penerbit LESTARI UKM, Bangi

Hashim HS, Aziz S, Abdul Aziz R (2011) Conservation with development: showcasing Langkawi Geopark. Plan Malays 1:1–24

Hashim HS, Abdul Aziz R, Aziz S, Ong PL, Chan KLG (2013) Pelan Pengurusan Langkawi Geopark (PPLG) 2012-2030. Penerbit LESTARI UKM, Bangi

Lee J (2012) Multilateral environmental agreements? A catalyst for interlinkages in river basin management? A case study of Pahang River Basin, Malaysia, river basins and change. Global Water System Project and UNESCO-IHE Institute for Water Education, Bonn, pp 188–193

Leman MS, Komoo I, Mohamed KR, Ali CA, Unjah T (2007) Geopark as an answer to geoheritage conservation in Malaysia—The Langkawi Geopark case study. Geol Soc Malays Bull 53:95–102

Mohamad AL, Ahmad N (2010) Conservation and biodiversity for the development of ecotourism in Langkawi Geopark. Akademika 30:31–38

Mohd Ayob N, Ahmad A, Mohamed B (2013) Spatial distributions of tourist in Langkawi Island. In: Proceedings of International conference on tourism development, pp 301–309

Mohd Yusof MF, Ismail HN, Raja Omar RN (2014) A critical analysis on evolution of branding destination in Langkawi Island. SHS Web of Conferences 12(01002):1–7

Murphy PE, Price GG (2005) Tourism and sustainable development, global tourism, 3rd edn. Elsevier (Butterworth-Heinemann), Amsterdam, pp 167–193

Naturally Langkawi (2017) Events Calendar 2017. http://naturallylangkawi.my/events/#1454953199998-05e7b38e-5ca3

Ong PL, Halim SA (2011) Planning for heritage tourism: the case of Langkawi Geopark. Plan Malays 1:121–144

Ong PL, Yusoff NH, Abdul Halim S (2010) Menjejak Semangat 'Kawi': Langkawi sebagai Geopark. Akademika 80:55–68

UNESCO (n.d.) Statutes of the International Geoscience and Geoparks Programme. http://www.unesco.org/new/fileadmin/MULTIMEDIA/HQ/SC/pdf/IGGP_IGCP_UGG_Statutes_Guidelines_EN.pdf

United Nations Educational Scientific and Cultural Organisation (UNESCO) (n.d.) Global Geoparks. Earth Sciences. Application Process for Aspiring UNESCO Global Geoparks. http://www.unesco.org/new/en/natural-sciences/environment/earth-sciences/unesco-global-geoparks/application-process/

United Nations General Assembly (UNGA), Resolution 53/243A (1999) Declaration on a Culture of Peace, UN Doc. A/RES/53/243A. http://www.un-documents.net/a53r243a.htm.

United Nations General Assembly (UNGA), Resolution 69/233 (2015) Promotion of Sustainable tourism, ecotourism, for poverty eradication and environmental protection, UN Doc. A/RES/69/233

United Nations World Tourism Organisation (UNWTO) (2011) Tourism—linking cultures: UNWTO launches world tourism day 2011 (PR No.: PR 11054) http://media.unwto.org/press-release/2011-06-20/tourism-linking-cultures-unwto-launches-world-tourism-day-2011

Var T, Ap J (1998) Tourism and world peace. In: Theobald WF (ed) Global tourism, 2nd edn. Butterworth Heinemann, Oxford, pp 44–61

World Travel and Tourism Council (WTTC) (2017) Travel and Tourism Economic Impact 2017 World. WTTC, London. https://www.wttc.org/-/media/files/reports/economic-impact-research/regions-2017/world2017.pdf

World Travel and Tourism Council (WTTC) (n.d.) Economic impact analysis. https://www.wttc.org/research/economic-research/economic-impact-analysis/

The Role of Ecotourism in Biocultural Landscape to Harmonize Nature and Human towards Sustainable Development: Clungup Mangrove Conservation Area as a Case Study

7

Rita Parmawati and Rizha Hardyansah

Abstract

Tourism development trends in the biocultural landscape have shown a significant increase, especially concerning tourist facilities. This has encouraged the emergence of problems between nature and humans. In several cases, many tourism destinations only highlight the attraction and neglect the efforts to preserve the environment. Meanwhile, the harmony between nature and humans provides many positive effects on the economic, agricultural, social, and tourism sectors. However, when this trend continues without harmony, it leads to the unsustainability of the environment. Therefore, to achieve sustainability for tourism in the biocultural landscape, the concept of ecotourism is pivotal. This is due to the symbiosis and relationship between nature and humans which serves as the main factors in the ecotourism concept. Moreover, ecotourism relies on the optimal use of natural and cultural landscapes and makes them attractive for tourism. Meanwhile, an example of the ecotourism application in the biocultural landscape was carried out by the communities in Clungup Beach, Malang Regency, namely Clungup Mangrove Conservation Area (CMC). Although the process still requires several optimization and development, its impact is felt directly by the community such as increased income, maintained natural conditions, and increased public awareness to protect the environment.

R. Parmawati (✉)
Socio-Economics Department, Faculty of Agriculture, Universitas Brawijaya, Malang, East Java, Indonesia

Postgraduate Program, Universitas Brawijaya, Malang, East Java, Indonesia
e-mail: rita_parmawati@ub.ac.id

R. Hardyansah
Master Program of Environmental and Development, Postgraduate, Universitas Brawijaya, Malang, East Java, Indonesia

S. A. Abdullah et al. (eds.), *Conserving Biocultural Landscapes in Malaysia and Indonesia for Sustainable Development*,
https://doi.org/10.1007/978-981-16-7243-9_7

Keywords

Biocultural landscape · Clungup Mangrove · Ecotourism · Nature conservation · Sustainable development

7.1 Introduction

Nature offers many benefits to humans, directly and indirectly, however, a gap appeared due to human desire for survival and the responsibility to preserve the environment. This gap has caused many conflicts in a society where some believe that the environment only exists to meet human needs. Moreover, humans have not realized that natural resources are divided into several types, namely renewable and non-renewable. Therefore, when nature is continuously used without considering its capacity, it causes a scarcity of natural resources.

One of the steps to preserve the environment is to limit the access and interaction between humans and natural resources. This causes conflicts between people that exploit natural resources and the government that formulated the policies. Moreover, the use of certain animal or plant species by communities for vital purposes has been suddenly restricted or completely prohibited due to the pursuit of environmental goals in the form of natural preservation. Although it has not led to harmony between humans and nature, it has the potential to trigger ongoing conflicts. Therefore, the fundamental question that arises from humans is whether environmental livelihoods can be eliminated for purposes such as biodiversity conservation, which causes a decrease in welfare and life opportunities, especially for rural communities (Veit and Benson 2004; Balisacan 2012). Previous studies have shown that a solution is needed to overcome this problem for humans and nature to be in harmony. In addition, the conservation of natural resources also makes the community prosperous.

Biocultural landscape tourism is one of the best choices to avoid conflicts of interest, preserve the natural environment, and increase the community's social and economic values (Kent 2003; Mensah and Ernest 2013; Blaj 2014). However, this does not generally apply to all types of tourism that are developed in a particular area due to several factors such as community support and tourists' awareness to maintain the surrounding ecosystem. In Indonesia, most of the types of tourism that are currently growing rapidly are mass tourisms that involve the arrival and gathering of many tourists in one destination. Moreover, during the holiday season, many sudden tourist destinations attract tourists to enjoy their vacation; however, when the holidays are over, these centers are often neglected. Meanwhile, improper management of mass tourism leads to a short-time increase in the economy of local communities and long-term ecosystem degradation around tourist destinations (Akis 2010; Lee and Syah 2018). Previous studies showed that there is a need to find alternative types of tourism development to achieve sustainability. Among the types of developing tourism that exists, ecotourism bridges the gaps that arise due to the need for environmental conservation which collides with the community's economic needs. Meanwhile, the concept of ecotourism appeared when the community realized the essential to create sustainability and balance between economic, ecological, and social (Diamantis 1999). According to experts, ecotourism is defined as part of tourism on natural resources that contributes to local conservation and community for tourists to gain new experiences and appreciate the local environment and culture (Ziffer 1989; Boo 1991; Australia Department of Tourism 1994; Figgis 1993; Boyd and Butler 1996; Lee and Jan 2019). Therefore, this concept is an important step to achieve sustainable development, especially in the biocultural landscape tourism sector (Buckley et al. 2008). Similarly, ecotourism also plays a role in developing the economy of people in rural areas including the biocultural landscape and increases local community awareness on nature conservation. However, ecotourism is not easily achieved because it requires synergy from national and local institutions.

Furthermore, the two other objectives of ecotourism apart from tourist experience include environmental conservation and sustainable development (Jamaliah and Powell 2018; McKercher 2010; Walter 2011, 2013). However, achieving ecotourism that improves the economy and maintaining environmental as well as socio-cultural sustainability has many challenges, especially in developing countries that are related to the strategic, planning, and operational sectors. In its application, tourists and community awareness are essential to maintain environmental sustainability for the future.

As one of the fastest-growing sectors of the global tourism industry, ecotourism is an environmentally, socio-cultural, and economically viable option to promote sustainable development in the biocultural landscape. Meanwhile, Schmitz and Herrero-Jauregui (2021) stated that the cultural landscape is used as a step to understand the socio-ecological dynamics of the community to formulate sustainable tourism development. Furthermore, cultural landscapes combined with ecotourism development protect the surrounding wilderness ecosystems. This feasibility makes ecotourism to play a significant role in harmonizing humans and nature which is the characteristic of the biocultural landscape. Moreover, the combination of tourism and landscape is complex because it involves many indicators such as social, economic, ecological, and policy (Jimenez-Garcia et al. 2020). Therefore, ecotourism is related to natural beauty and the local culture of the community. Also, it synergizes human economic needs and for nature conservation. However, there is a possibility of conservation risks due to tourist arrivals and pollution (Zhang et al. 2012; Zawilinska 2020). In addition, it is one of the several attempts to reconcile humans' fundamental rights with environmental needs. One of the tourist destinations developing ecotourism is Clungup Mangrove Conservation (CMC). Meanwhile, the existence of this process has a multiplier effect on the economic sector (Dharma et al. 2021), contributes to mangrove conservation and rehabilitation (Harahab and Setiawan 2017; Hakim et al. 2017), and has demonstrated sufficient support for conservation-based sustainable development (Wibowo et al. 2018). Therefore, this chapter aims to determine the role of ecotourism in creating a balance between humans and the environment in the biocultural landscape using Clungup Mangrove Conservation (CMC), in Malang Regency, East Java as a case study.

7.2 Nature and Humans: Towards Sustainable Development

Nature and humans have a close relationship which is reflected in the behavior of people using natural resources for economic benefits through reproduction. Currently, human interaction with nature had not occurred directly. However, with the industrial revolution that occurred in 1750–1850, nature has to be beneficial in various forms as previously discussed. Fuller and Irvine (2010) and Irivne et al. (2010) stated that after World War II and industrialization, human interactions began to change from the desire for direct consumption and exploitation to become mutualistic interactions such as tourism. Meanwhile, interaction with nature is also essential to improve the quality of human life (Aldous 2007; Bowler et al. 2010). Some of the benefits from human interaction with nature include improved physical health (Ulrich 1984; Maas et al. 2006; Richardson and Mitchell 2010), which have a positive impact on psychological conditions (Bodin and Hartig 2001; Kaplan 2001; Fuller et al. 2007), increase the ability to receive information and experience (Han 2009), and create harmonious or relationships with humans (Shinew et al. 2004). Therefore, interacting with nature directly or indirectly improves humans' welfare, whose growth rates are higher.

Based on previous explanations, interactions with nature have a positive impact on humans. The benefits of mini-style interaction are seen when children play more often in nature and pay attention to their surroundings. Therefore, this enables the development of behavior to protect the environment and increase the ability to think and socialize (Wells 2000; Kuo and Sullivan 2001). A previous study conducted by Lohr and Pearson-Mims (2005) stated that children's interactions with nature affect their attitudes towards the future environment and increase the potential for environmental conservation efforts. Similarly, studying the positive impacts of human interactions with nature also fosters a perspective of sustainability. With a positive relationship between humans and nature, activities such as exploitation trigger the appearance of humans' rejection, because it leads to ecosystem degradation (Millennium Ecosystem Assessment 2005). Currently, nature has been degraded by several activities such as agricultural intensification, deforestation, pollution, and land disturbance, which reduce the quality of human life and threaten the future. The solutions to overcome this are also limited. However, large-scale conservation, effective land management, and prohibiting activities degrade the environment. Moreover, community participation is also required to be in line with environmental conservation goals. Therefore, to increase community participation, an attraction from the conservation activities needs to be carried out. Furthermore, integration of the concept of sustainable tourism, such as ecotourism, is a suitable alternative to accelerate conservation and effective land management without reducing the economic potential of people that depend on nature. This process is essential because protecting and enhancing biodiversity is important for a sustainable future and improve life quality in the current ecosystem transition period.

7.3 Ecotourism in Clungup Mangrove Conservation (CMC)

Clungup Mangrove Conservation (CMC) is one of Indonesia's ecotourism objects, which is used as a reference point for several natural tourism managers to develop into ecotourism. Geographically, CMC is located in Tambakrejo Village, Sumbermanjing Wetan District, Malang Regency, East Java (Fig. 7.1).

CMC was initiated by the Bhakti Alam Sendang Biru Foundation because of concern over the degraded natural resources in South Malang, especially around Clungup Beach. In 2012, the natural conditions around the coast were damaged due to unfriendly uses. This situation began in 1999 when the community started illegal felling of trees, for the sale of mangroves, and converting to fish ponds using non-environmentally friendly methods such as potassium extensification of forest land, specially protected forests, into agricultural land.

Moreover, CMC has an area of more than 81 hectares designated as a mangrove forest area and, a coastal border area or greenbelt of 117 hectares (Harahab and Setiawan 2017; Husamah and Hudha 2018). Based on the Department of Fisheries and Marine Affairs of East Java Province, there are several zones in CMC, which are shown in Table 7.1.

This rehabilitation activity was directed and synergized with the concept of tourism for Yayasan Bhakti Alam Sendang Biru branded CMC as a mangrove conservation ecotourism destination. Moreover, the ecotourism area consists of Clungup Beach, Gatra Beach, and Tiga Warna Beach. In 2012, these communities

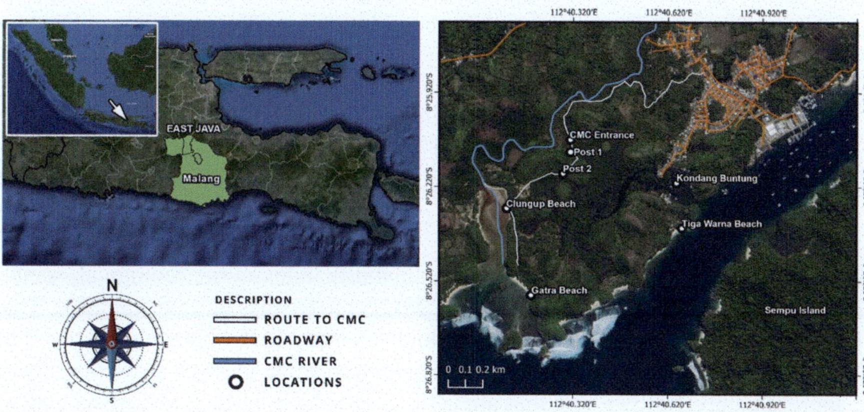

Fig. 7.1 Clungup Mangrove Conservation area

Table 7.1 Zoning in Clungup Mangrove Conservation (CMC)

No	Zone	Name of the coastal
1	Core Zone	Pesisir Pantai Tiga Warna
2	Rehabilitation Zone	Pesisir Pantai Tiga Warna
3	Mangrove	Pesisir Desa Sitiarjo
4	Mangrove Core Zone	Kondang Buntung

succeeded in developing the CMC area as an ecotourism destination, which later became a reference by several communities, especially in coastal areas. Furthermore, the CMC ecotourism area won the Adi Bakti Mina Bahari national competition in 2015 from the Ministry of Marine Affairs and Fisheries because it represented balanced tourism management between humans and the environment towards nature conservation and sustainability. In every activity, CMC management is always consistent with promoting and providing information related to the importance of mangroves in coastal ecosystems (Hakim et al. 2017).

The Ecotourism CMC management by Yayasan Bhakti Alam Sendang Biru has several regulations that need to be followed to visiting tourist destinations. These regulations aim to make nature conservation activities in the CMC area more organized and moving constantly. In addition, these regulations considered the ecosystem's carrying capacity and aimed to educate tourists for further visitation. Therefore, tourists gain more knowledge related to mangrove conservation in coastal areas. Some of the existing regulations include:

1. Limiting the number of visitors and time to visit, which is made to reduce the negative impact of tourist arrivals on the ecosystem.
2. Luggage inspection system, tourists' luggage is always checked to avoid the bringing of potentially harmful items to the environment. Furthermore, trashed items are also recorded and checked when leaving the area, which aims to prevent tourists from littering in the ecotourism area.
3. The obligation to participate in nature preservation, CMC management requires tourists to pay an entrance ticket with a value similar to the price of one mangrove tree seedling. Also, there is a fee that makes tourists plant mangroves in the area around CMC.
4. There is a system of opening and closing the ecotourism area, which is usually referred to as ecosystem restoration. This closed activity aims to make nature have time to regenerate its ecosystem without any disturbance from humans (tourists), and the CMC management carries out tracing and cleans tourist destinations from the rubbish.

Wibowo et al. (2018) stated that the sustainability index of the ecological, social, economic, and institutional dimensions in CMC ecotourism is quite sustainable. This is due to the active role of the six main volunteers from CMC that were involved in the process of ecosystem conservation and land restoration in 2012. Presently, there are approximately 108 workers that are actively involved in the CMC management process. Moreover, CMC has implemented a reservation system, construction of monitoring posts, and restrictions. The two monitoring posts, namely a tourist registration and a checklist of items that have the potential to become waste. Meanwhile, the second is the recording of waste taken out where the number of waste items that have been recorded in the first is equal to the amount checked. Moreover, violations by tourists require strict sanctions. In addition, the restrictions imposed by the management are on the number of tourists and the time of visit. The total number of tourists allowed is approximately 700 people with details on Tiga

Warna Beach which is limited to 100 for a maximum of 3 h with the obligation to use local guide services. On the Clungup and Gatra Beaches, only 300 people are allowed on each beach and a maximum of 75 tents, when the tourist desires to camp. Therefore, CMC management has applied conservation principles in every process to maintain the sustainability of the existing mangrove ecosystem.

Meanwhile, the opportunities that exist due to the excellent management of this CMC have not been fully utilized by the surrounding community. Based on Husamah and Hudha (2018), managers are still unable to provide empowerment or training to the community to be involved as economic actors in the CMC area. In line with that study, Harahab et al. (2020) stated that managers still focused on the process of rehabilitation and conservation of degraded land. This requires very large funds. Therefore, the process of community empowerment remains an agenda that has not been realized by CMC managers. Furthermore, the conservation and rehabilitation process shows a significant change in the recorded mangrove cover from 2010 to 2016 which increased from 3015–4131 ha or by 0.828 ha (Wonorengga 2017). This also plays an active role for CMC managers in remaining obedient to the conservation-based policies that have been made. A previous study by Dharma et al. (2021) showed that the daily carrying capacity of the CMC is estimated to accommodate approximately 3000 people per day. However, due to the firmness and compliance of the management, the number of visitors and the number of visits for the CMC area is limited to maintain the stability of the surrounding ecosystem.

7.4 Discussion and Conclusion

One form of harmonization between humans and nature is creating tourist destinations that use their natural potential by minimizing damage and maximizing conservation programs as well as community welfare. Meanwhile, the combination of coastal nature tourism and mangrove forest conservation activities in the biocultural landscape of CMC Ecotourism is an example of this success. A previous study by Hakim et al. (2017) and, Imron and Anwar (2019), stated that CMC is classified as tourism that uses mangroves to increase economic value and carry out conservation activities in degraded areas.

Ecotourism activities have a significant impact on the increase of the harmony between nature and humans in the biocultural landscape. Meanwhile, these activities also have an impact on improving the quality of the surrounding ecosystem, local wisdom preservation, and tourist satisfaction (Koki 2017) which also improves and develops the growth of the local economy (ILO 2012). Other tourism activities oriented towards local communities are relatively few, while many have failed in the development process due to a lack of balance, such as burdening only one activity (either tourism or economy only). When there is no balance, the harmony of nature and humans is difficult to create because humans make use of nature's potential without thinking about its sustainability. Meanwhile, the basic concept of ecotourism is to encourage tourists and local communities to better understand the environment by actively engaging tourism activities in the conservation of flora, fauna, ecosystems, and land. Although many relationships need to be created, the essential link in the development and management of ecotourism is the physical aspect and the biological landscape (Beeton 2001; Fisher et al. 2020).

As the initiator of CMC, Yayasan Bhakti Alam Sendang Biru has a guideline of *Living with Nature/Hidup Bersama Alam,* which is adopted from the *Tri Hita Karana* concept. Moreover, *Tri Hita Karana* is a concept and philosophy of Hindus' life, which means "*Three Causes of Happiness.*" This teaching emphasizes the relationship between humans, nature, and God. In its implementation, these three relationships need to be balanced and harmonized to create harmony and balance of life; therefore, humans have an important (Puspitadewi et al. 2015; Parmajaya 2018). Similarly, guidelines for Living with Nature make CMC a definition that in its development seeks to use natural resources in a balanced or sufficient manner. These guidelines are also promoted to the surrounding communities in Tambakharjo Village. This makes the CMC management activities empower local communities for nature exploitation, such as illegal logging and extensification of agricultural land to be reduced or stopped.

Observations have shown that the growth in human and nature's needs to survive is becoming a problem that requires a fair settlement for both. Therefore, harmony between humans and the environment is essential for sustainability. This means that humans use nature wisely and does not lead to resource exploitation to maintain natural conditions and capacities for a long time. As a provider of human needs, nature also benefits through conservation activities to create a sustainable environment. Meanwhile, economic needs and environmental sustainability are two sides that constantly collide after completion. Therefore, with the presence of the ecotourism concept at CMC, there has been an increase in the quality of the ecosystem indicated by a significant distribution of aquatic bird species (Rachmaputra et al. 2018). This is because the area of mangrove land cover is increasing every year (Wonorengga 2017) and also increases the carrying capacity of the environment which affects the economy of the surrounding community. From an economic point of view, it is considered less optimal because people are unable to read the existing opportunities. However, the CMC management continues to involve and prioritize the surrounding community to support ecotourism activities by building open homestays, stalls, rest areas, or offering taxi services to the registration location. Currently, there are still many improvements in the quality of services and cooperation between local communities and managers. Based on these improvements, it is believed that ecotourism becomes a bridge capable of creating harmony between nature and humans. Therefore, the behavior to exploit nature has turned into a desire to coexist with nature.

Acknowledgments The authors are grateful to Postgraduate Brawijaya University and Conservation Activists from Clungup Mangrove Conservation (CMC) Malang Regency who have supported and helped us in compiling the data of this research.

References

Akis A (2010) The effects of mass tourism: a case study from Manavgat (Antalya-Turkey). Procedia Soc Behav Sci 19:289–296

Aldous DE (2007) Social, environmental, economic, and health benefits of green spaces. Acta Hortic 762:171–184

Australia Department of Tourism (1994) National ecotourism strategy. Australia Government Publishing Service, Canberra

Balisacan RHC (2012) Harmonizing biodiversity conservation and the human right to livelihood: towards a viable model for sustainable community based ecotourism using lessons from the Donsol Whale Shark Project. Anteneo Law J 57:423–462

Beeton S (2001) Ecotourism: impacts, potentials and possibilities. Tour Manage 22(5):576–578

Blaj R (2014) Ecotourism and nature tourism – components of a sustainable management of forests. J Hortic For Biotechnol 18(4):51–54

Bodin M, Hartig T (2001) Does the outdoor environment matter for psychological restoration gained through running? Psychol Sport Exerc 4:141–153

Boo E (1991) Planning for ecotourism. Parks 2(3):4–8

Bowler DE, Buyung-Ali LM, Knight TM, Pullin AS (2010) A systematic review of evidence for the added benefits to health of exposure to natural environments. BMC Public Health 10:456

Boyd SW, Butler RW (1996) Managing ecotourism: An opportunity spectrum approach. Tour Manage 17(8):557–566

Buckley R, Cater C, Linsheng Z, Chen T (2008) Shengtai Iuyou: cross-cultural comparison in ecotourism. Ann Tour Res 35(4):945–968

Dharma PA, Yulianda F, Yulianto G (2021) Suitability and carrying capacity of coastal ecotourism in Clungup Mangrove Conservation (CMC), Malang District, East Java. ECSOFiM 8(2):196–210

Diamantis D (1999) The concepts of ecotourism: evolution and trends. Curr Issue Tour 2 (2–3):93–122

Figgis PJ (1993) Ecotourism: Special interest or major direction? Habitat Australia 21(1):8–11

Fisher MR, Verheijen B, Sahide MAK (2020) Community and conservation in Wallacea: Making the case for the region, a methodological framework, and research trends. For Soc 4:1–19

Fuller RA, Irvine KN (2010) Interactions between people and nature in urban environments. In: Gaston KJ (ed) Urban ecology. Cambridge University Press, Cambridge

Fuller RA, Irvine KN, Devine-Wright P, Warren PH, Gaston KJ (2007) Psychological benefits of greenspace increase with biodiversity. Biol Lett 3:390–394

Hakim L, Siswanto D, Nakagoshi N (2017) Mangrove conservation in east java: the ecotourism development prospectives. J Trop Life Sci 7(3):277–285

Han K (2009) Influence of limitedly visible leafy indoor plants on the psychology, behaviour, and health of students at a junior high school in Taiwan. Environ Behav 41:658–692

Harahab N, Setiawan (2017) Suitability index of mangrove ecotourism in Malang Regency. ECSOFiM 4(2):153–165

Harahab N, Riniwati H, Zulfaidah CA (2020) Analisis Partisipasi Masyarakat dalam Pengelolaan Ekowisata Clungup Mangrove Conservation (CMC). J Fish Mar Res 4(2):296–307

Husamah, Hudha AM (2018) Evaluasi Implementasi Prinsip Ekowisata Berbasis Masyarkat dalam Pengelolaan Clungup Mangrove Conservation Sumbermanjing Wetan, Malang. Jurnal Pengelolaan Sumberdaya Alam dan Lingkungan 8(1):86–95

ILO (2012) Rencana Strategis Pariwisata Berkelanjutan dan Green Jobs untuk Indonesia. Jakarta: Kementerian Pariwisata dan Ekonomi Kreatif Republik Indonesia bekerjasama dengan International Labour Organization (ILO)

Imron M, Anwar MS (2019) Strategi Kolaborasi Pengembangan Wisata Berbasis Edukasi di Clungup Mangrove Conservation Desa Tambakrejo Kabupaten Malang. J Gov Innov 1(1):78–90

Irivne KN, Fuller RA, Devine-Wright P, Payne S, Tratalos J, Warren P, Lomas KJ, Gaston KJ (2010) Ecological and psychological value of urban green space. In: Jenks J, Jones J (eds) Dimensions of the sustainable city. Springer, Dordrecht

Jamaliah MM, Powell RB (2018) Ecotourism resilience to climate change in Dana Biosphere Reserve, Jordan. J Sustain Tour 26(4):519–536

Jimenez-Garcia M, Ruiz-Chico J, Pena-Sanchez AR (2020) Landscape and tourism: evolution of research topics. Land 9(488):1–17

Kaplan R (2001) The nature of the view from home: psychological benefits. Environ Behav 33: 507–542

Kent M (2003) Ecotourism, environmental preservation and conflicts over natural resources. Horizontes Antropológicos 9(20):185–203

Koki JN (2017) Contribution of ecotourism towards sustainable livelihood of the communities living on Wasini Island, Kwale County, Kenya. J Tour Hospit Manage 5(3):106–125

Kuo FE, Sullivan WC (2001) Aggression and violence in the inner city: effects of environment via mental fatigue. Environ Behav 33:543–571

Lee TH, Jan FH (2019) Can community-based tourism contribute to sustainable development? Evidence from residents' perceptions of the sustainability. Tour Manag 70:368–380

Lee JW, Syah AM (2018) Economic and environmental impacts of mass tourism on regional tourism destinations in Indonesia. J Asian Fin Econ Bus 5(3):31–41

Lohr VI, Pearson-Mims CH (2005) Children's active and passive interactions with plants influence their attitudes and actions toward trees and gardening as adults. Hort Technol 15(3):472–476

Maas J, Verheuj RA, Groenewegen PP, de Vries S, Spreeuwenberg P (2006) Green space, urbanity, and health; how string is the relation? J Epidemiol Commun Health 60:587–592

McKercher B (2010) Academia and the evolution of ecotourism. Tour Recr Res 35(1):15–26

Mensah I, Ernest A (2013) Community participation in ecotourism: the case of Bobiri Forest Reserve and butterfly sanctuary in Ashanti Region of Ghana. Am J Tour Manage 2(1A):34–42

Parmajaya IPG (2018) Implementasi Konsep Tri Hita Karana dalam Perspektif Kehidupan Global: Berpikir Global Berperilaku Lokal. Purwadita: Jurnal Agama dan Budaya 2(2):27–33

Puspitadewi IDA, Windia W, Astiti NWS (2015) Analisis Pengaruh Penerapan Tri Hita Karana terhadap Tingkat Kepuasan Wisatawan di Kawasan Agrowisata Ceking, Kecamatan Tegallalang, Kabupaten Gianyar. Jurnal Manajemen Agribisnis 3(1):43–52

Rachmaputra A, Yanuwiadi B, Leksono AS (2018) Species diversity of birds in Clungup Mangrove Conservation Sendang Biru, Kabupaten Malang, East Java as bioindicator. J Environ Eng Sust Technol 5(2):47–51

Richardson EA, Mitchell R (2010) Gender differences in relationships between urban green space and health in the United Kingdom. Soc Sci Med 71:568–575

Schmitz FM, Herrero-Jauregui C (2021) Cultural landscape preservation and social-ecological sustainability. Sustainability 13(2593):1–5

Shinew KJ, Glover TD, Parry DC (2004) Leisure spaces as potential sites for interracial interaction: community gardens in urban areas. J Leisure Res 36:336–355

Ulrich RS (1984) View through a window may influence recovery from surgery. Science 224:420–421

Veit PG, Benson C (2004) When parks and people collide. Environmental rights. Spring

Walter P (2011) Gender analysis in community-based ecotourism. Tour Recr Res 36(2):159–168

Walter PG (2013) Theorising visitor learning in ecotourism. J Ecotour 12(1):15–32

Wells NM (2000) At home with nature: effects of "greenness" on children's cognitive functioning. Environ Behav 32:775–795

Wibowo FAA, Riniwati H, Nugroho I (2018) the role of volunteer for the management of conservation-based ecotoruism in Clungup Mangrove Conservation Tambakrejo Village, Sumbermanjing Wtan Sub-District, Malang. J Indones Tour Dev Stud 6(3):187–193

Wonorengga Y (2017) Analisis spasial dan temporal perubahan tutupan mangrove di kawasan Clungup Mangrove Conservation Bhakti Alam Sendang Biru. Faculty of Fisheries and Marine Sciences, Brawijaya University, Malang, Kabupaten Malang. Thesis

Zawilinska B (2020) Residents' attitudes towards a national park under conditions of suburbanisation and tourism pressure: A case study of Ojcow National Park (Poland). Eur Countrys 1:119–137

Zhang JT, Xiang CL, Li M (2012) Integrative ecological sensitivity (IES) applied to assessment of eco-tourism impact on forest vegetation landscape: A case from the Baihua Mountain Reserve of Beijing. China Ecol Indic 18:365–370

Ziffer KA (1989) Ecotourism: the uneasy alliance. Conservation International and Ernst & Young, Washington

Understanding Micro-experiences of Heritage Conservation in an Island-based Tourism Development: A Case of Kubang Badak BioGeoTrail, Langkawi UNESCO Global Geopark, Kedah, Malaysia

8

Sharina Abdul Halim, Tanot Unjah, Norhayati Ahmad, and Ibrahim Komoo

Abstract

Around the world, efforts to facilitate, promote and strengthen heritage conservation initiatives in an island-based tourism development have been widely carried out. Still, more works need to be conducted to truly achieve its goals. One of the main issues facing heritage conservation is the lack of holistic and integration to ensure participation from the local community and stakeholders in the development process cycle. The question arises: how is combining local knowledge and values assists conservation? Thus, using Kubang Badak BioGeoTrail as an example, this chapter attempts to address this question by examining micro-experiences of the local community living in heritage areas with rapid tourism development activities. Micro-experiences provide unique insights into diverse ways people are embedded in their environments based on their knowledge and responses. Therefore, we will be able to understand local community livelihoods, social relations and cultures that are most intimately linked to the environment, particularly in heritage conservation per se. Langkawi Island has been receiving more than two million visitors annually since its inception as a duty free status in

S. A. Halim (✉) · T. Unjah · I. Komoo
Institute for Environment and Development (LESTARI), Universiti Kebangsaan Malaysia, Bangi, Selangor Darul Ehsan, Malaysia
e-mail: sharinahalim@ukm.edu.my; tanot-u@ukm.edu.my; ikomoo@ukm.edu.my

N. Ahmad
Department Biological Sciences and Biotechnology (Biology Programme), Faculty Science and Technology, Universiti Kebangsaan Malaysia, Bangi, Selangor Darul Ehsan, Malaysia

Present Address: Institute for Environment and Development (LESTARI), Universiti Kebangsaan Malaysia, Bangi, Selangor Darul Ehsan, Malaysia
e-mail: norhayatiahmad@ukm.edu.my

S. A. Abdullah et al. (eds.), *Conserving Biocultural Landscapes in Malaysia and Indonesia for Sustainable Development*,
https://doi.org/10.1007/978-981-16-7243-9_8

late 1980s. Tourism posed opportunities and various threats related to sustainability matters of the island from the environmental, economic, physical, social and cultural aspects. One of the concerted efforts undertaken to address these issues is through the establishment of Langkawi UNESCO Global Geopark in 2007, which strives to balance conservation with development activities, in order to fulfil the needs of socio-economic, well-being and education for the whole community. Careful analysis and appreciation of micro-experiences of local individuals and communities may therefore broaden the foundation for shared understanding and collaborative action to address sustainability issues, particularly in balancing heritage conservation and development inclusively and effectively.

Keywords

Geopark · Integrated heritage conservation · Livelihoods · Micro-experiences · Participatory · Sustainability

8.1 Introduction

The challenge to have both conservation and development in efforts to protect resources that are of heritage significance is still much debated particularly in developing countries (Hashim et al. 2011; Hes 2017; Bennett et al. 2017; Reed et al. 2019). Communities, practitioners, scholars, interest groups and agencies have been challenged to negotiate rather than "defend" certain notions and thus avoid the traditional conflict between conservation ideas on one side and development on the other. Heritage conservation requires a participatory approach from the very start of the process, namely planning, implementing, monitoring and evaluating. As iterated by Sacco et al. (2013), heritage-led development should be based on negotiation between the economic and social conditions that enable the heritage to be performed and executed through the proper strategic coordination with the local economy and community.

This way development is more in tune with its expectations (Mahjabeen et al. 2009). More importantly, it is not possible to plan sustainable heritage development without taking into consideration the cultural processes at the everyday, micro-level (Ilmonen 2015). As stated by Nasser (2003), the critical elements in promoting synergy between heritage and development are four-fold:

- The need for long-term planning (local community and stakeholders are involved in setting long-terms goals, but with short-term steps at the local level).
- The need to protect heritage, since it will be degraded if over-exploited.
- Change and development are necessary.
- Equitable access to heritage and culture resources by the local community and stakeholder.

In a broader context, local culture is an essential component of regional development, and its protection and development must be closely integrated with planning and developing the entire region. Before any new development is initiated, the local culture must be evaluated and a great deal of knowledge, creativity and innovation must be invested into it to gain the development benefits for the community (Nared and Bole 2020). Having constructive engagements with local communities and all stakeholders are crucial to streamline their participation in heritage conservation and construction of narratives (Adom 2019).

8.2 Micro-experiences and Participatory Approach Towards Sustainable Heritage Conservation

In this chapter, we focus on the importance of understanding how species and landscape influences the way we humans interact constantly, illuminated through micro-experiences (Burke et al. 2020). It is how individuals and communities are experiencing changes in the environment, how they construct knowledge based on these experiences and how that knowledge shapes their responses. Examining heritage conservation at the resolution of micro-experiences has the advantage of showing us change where many people, especially for those whose livelihoods, social relations and cultures are most intimately linked to the environment.

Micro-experiences in this chapter also refers to the subjective and humanistic dimension of sustainability that acknowledges and respects local voices, values and perceptions to foster social learning and collaboration with other stakeholders in solving complex problems to attain more sustainable heritage conservation and management. In essence, the significance of micro-experiences is inspired by Ostrom's breakthrough work in *Governing the Commons* (1990) where self-organising solutions to commons problems are not only possible but also occur quite often regarding sustainable use of shared resources. Communities can solve common dilemmas without complex formal infrastructures. Instead, they rely on a combination of informal norms, trust and a small set of formal rules that the users themselves construct, monitor and enforce. Among many other significant contributions, Ostrom's also forged essential linkages between the social and natural sciences.

In recent decades, international discourse on heritage conservation has solicited a closer relationship between cultural, social and economic issues and sustainable development, giving rise to an interest using natural resources and cultural capital as bearers of evolutionary potential for human societies. As Ronchi et al. (2020) eloquently described the analogy between the role of cultural diversity for social development and biodiversity for the evolution of the natural ecosystem, "one can conclude that conservation and the transformation of cultural heritage may be considered an ethical duty, according to the principles of intergenerational and intragenerational equity". Thus, conservation of natural resources produced collective benefits in local areas through socio-economic activities that are sustainable (Adom 2019). Evidently, if nature-based tourism development (i.e. ecotourism and geotourism) would flourish in local regions, active participation of local people and their cultural tradition is imperative (Fuller et al. 2007; United Nations 2003; Adom 2018; Zhang et al. 2020).

Several typologies of participation distinguish between processes of participation depending on the scope of public involvement, with lower (less participatory ends of the spectrum including information and consultation) and higher (more participatory) ends incorporating approaches that seek collaboration "two-way exchange" and empowerment of participants (i.e. Arnstein 1969; Pretty 1995). It is also crucial that the scope of public participation process is made explicit from the outset, to avoid raising false expectations (Conrad et al. 2011). The relevance of participation to heritage conservation is evident if the landscape is understood to be a product of the interaction between nature and people. However, it is also challenging to ensure effective participation due to various factors, such as limit of top-down regulation, limited influence of the public on outputs and gap between policies and practices at the operational level (Conrad et al. 2011).

Acknowledging these challenges, success in any heritage conservation initiatives would still require a participatory approach, to facilitate this coordination of heritage conservation. The participatory process is not only to gain new knowledge but also to assist local communities, agencies, researchers and practitioners with open debates about developmental and societal issues (Nared and Bole 2020; Chitty 2017). It also allows local people to go through that process of encoding their cultural memories and history of their landscape (Shea 2019). Reproducing and recalling cultural memory is place-specific, and landscapes are vessels for family stories and community memories (Li 2010). Thus, the link formed between cultural memories, everyday activities and landscapes, are essential in forming a clear sense of the past and reinforcing the attachment to places in the present as well as future context.

There is an interconnectedness between the cultural and physical realms. In the Solomon Islands, for example, archaeological remains that are scattered through the rainforests of Solomon Islands are more than static remnants of history, as said by Walter and Hamilton (2014),

> When an Isabel Islander walks through the forest, moving between named sites and places, history is revealed and the journey helps structure or reinforce individual and group identity

The landscape is part of the lived world of the islanders. It symbolises the recognition of the embedded nature of the relationship between the cultural and physical domain of land (Walter and Hamilton 2014). In sum, this landscape becomes part of the islanders' cultural and natural heritage.

Such importance of the landscape as a source of individual and group identity is similarly experienced in the island of Langkawi, Kedah, Malaysia (Tan and Bouchaud 2015). Although the significance of nature and culture in Langkawi Island is unique, both elements are essential as part and parcel of the everyday lives of the

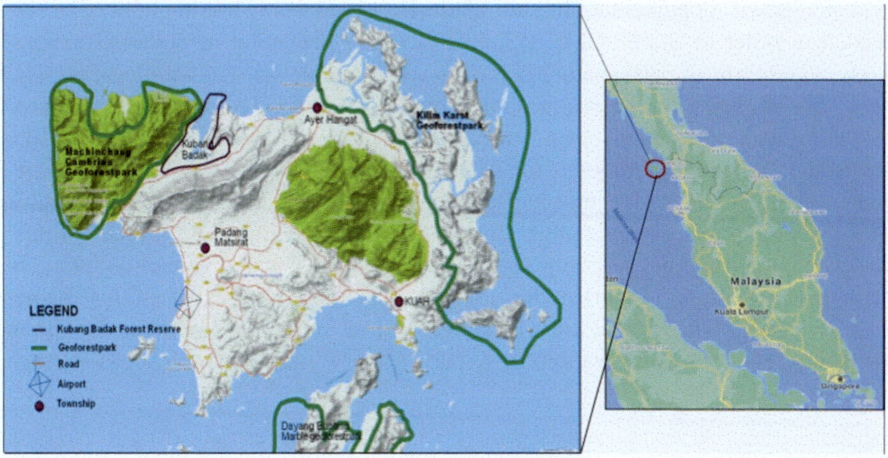

Fig. 8.1 Location of Kubang Badak BioGeoTrail, Langkawi, Kedah, Malaysia (source: Komoo et al. 2018)

local people living harmoniously with nature. The various accolades received from the Island of Legends, 99 Magical Islands, Duty Free Island, to Naturally Langkawi, among many, are acknowledgements that Langkawi is simply an innumerable beauty, as described by the *Sanskrit* word, Langka means innumerable and Wi means beauty (see Fig. 8.1).

The archipelago features distinctive geology and biodiversity with complete sedimentary sequence dating from 541 million to 252 million years ago and the oldest rock formations in the whole of Malaysia (Leman et al. 2007a, b). In 2007, UNESCO bestowed on Langkawi the status of Global Geopark, officially recognising her importance as a unique geological site. However, having geopark is not all about rocks, more importantly geopark promotes integrated heritage conservation together with empowering local communities in a participatory manner through innovative and sustainable socio-economic development (Komoo and Patzak 2008).

One of the main objectives of a geopark is to stimulate economic activity within the framework of sustainable development. Respectful of the environment, the creation of innovative local enterprises, small businesses, cottage industries initiates high quality training courses and new jobs by generating new sources of revenue (i.e. geotourism and geoproducts) (GGN 2010). Geotourism is an economic, success-oriented and fast-moving discipline, a new tourist business sector involving strong multidisciplinary cooperation (Komoo et al. 2018). Importance of conserving heritage resources in an integrated manner would boost geotourism activities that promote features consisting of geological monuments (geosites and geoareas), biodiversity, socio-cultural heritage (history, legend and myths) holistically.

Even though, a wealth of literature about geotourism exists, there still appears to be very little research focusing on the visitors and local communities of geopark territories (Gonzalez-Tejada et al. 2017; Halim et al. 2011; Farsani et al. 2011). More evidence-based documentation and platform to encourage communication are needed in order to share and shed light of experts managing resources collaboratively, via a form of participatory democracy or a citizen science with local population. This had led to recurrent questions about how do local knowledge and shared values assist conservation? In a sense, how local communities can meaningfully contribute to the interpretation at the in situ (e.g. guides, demonstrations) and ex situ context (e.g. gallery centre, museums, educational materials). Using Kubang Badak BioGeoTrail as an example, this chapter explores the questions posed through iterative learning process of micro-experiences from local community living in heritage area with rapid tourism development. The modest objective of this chapter is to ascertain whether local knowledge and shared values are important drivers in managing heritage conservation in a participatory manner through geotourism initiative.

8.3 Kampung Kubang Badak, Langkawi, Kedah, Malaysia

Kampung (Kg.= village) Kubang Badak has a unique ecosystem, representing a river estuary influenced by the tides, endowed with high geological diversity, pristine mangrove forests and a history of the early settlement of Thai-origin communities (see Fig. 8.1). The wetland area in Kubang Badak is rather small, estimated about 1000 ha, of which 514.3 ha comprise of the Kubang Badak Forest Reserve (Majlis Perbandaran Langkawi Bandaraya Pelancongan 2002). Areas outside the forest reserve consists mainly of rice paddy (now mostly abandoned and used for cow and buffalo grazing), paper bark woodland, inter-tidal flats, secondary forest on steep slopes and some original forest on the limestone forest (Wetlands International Malaysia 2005). The main land use in the area adjacent to the wetlands used to be related to active rice farming and cattle grazing although there is a large cement mining operation in the limestone karst to the east of Tg. Mendidih.

There are three main villages in the area consisting of Kg. Kubang Badak, Kg. Kelubi and Kg. Ewa. The population of Kg. Kubang Badak is approximately 1000 with about 93% ethnically Malay (Interview with Chairman of Fishermen Co-operative of Sungai Kubang Badak). The Fishermen Co-operative Group of Sg. Kubang Badak, a community-based organisation under the provision of Department of Fisheries Malaysia, consists of a total of 118 members, out of which 30 are full time members that are fisherfolks and the rest of them are part time members who are working in various sectors, such as tourism, hospitality and services, government and industry. Most of the villagers in Kubang Badak have diversified their livelihoods, from primarily fishing to service-based activities, due to its strategic location in nearby tourism landmarks and high-end resorts and accommodations.

There have been few previous studies in this area although the site has been identified as an important area for fisheries and ecotourism development (Wetlands International Malaysia 2005; Bolwerk 2018). Based on the *Rancangan Tempatan*

Daerah Langkawi (RTDL) 2030, the unique limestone karst and mangrove forest landscape in Kubang Badak Forest Reserve as well as its endemic flora, fauna and overall ecology are highlighted. There are no specific development plans stated for the area, but it is stated that the Kubang Badak Forest Reserve shall remain as such and only be promoted for passive eco- and geotourism activities (RTDL 2030).

Major threats to the ecological integrity of the system were considered low, although there were several identified threats, such as expansion of the limestone quarrying, introduction of alien species, and the escape crocodiles from the nearby Langkawi Crocodile Farm (Wetlands International Malaysia 2005).

The future of the Kubang Badak estuary appears to be safe-guarded through the stated objectives of the Langkawi District Local Plan (2030). However, there is still a need for close coordination between several sectors to ensure that the approach taken is sustainable. Most notably, any expansion or planned expansion of the YTL Teluk Ewa limestone quarry and cement factory will need to be carefully assessed in relation to the environmental, health and social impacts and local and state government investments to promote the tourism sector at Kubang Badak. Appropriate mitigation and adaptation measures need to be in place to conserve the Kubang Badak ecological integrity and high landscape value from potential impact of mass industrial and tourism development in the future.

As such, the development of Kubang Badak BioGeoTrail (Fig. 8.1) is a strategic partnership project between the Langkawi Development Authority (LADA), Langkawi Research Centre (PPL), Universiti Kebangsaan Malaysia and Fishermen Cooperative of Sungai Kubang Badak, Langkawi, to produce a new geotourism or knowledge-based tourism product in order to achieve maximum satisfaction among tourists and to inculcate the idea of conservation and management of natural resources in a sustainable way among local communities and visitors. The emphasis of biological and geological elements (BioGeo) in naming the trail is based on unique human interdependence with natural environment situated in a permanent forest reserve with pristine mangrove ecosystem as well as outstanding geological significance related to the oldest rock formation of Malaysia, Machinchang Cambrian (550 million years old).

Geotourism development in a geopark focuses on balancing efforts in conserving geological heritage and landscapes, introducing links between natural and cultural heritage enhancing the socio-economics of local communities. This is a concerted effort undertaken since 2015 until 2019, to raise public awareness and promote new geotourism initiative with interesting findings that link geological, biological and cultural heritage at each site of the trail. The development of Kubang Badak BioGeoTrail is not for conservation per se but ultimately strives to pursue its potential to generate the local economy of the inhabitants, who are primarily involved in fishing, agriculture, hospitality and service sectors. The opportunity of this trail development is also a window to reviving back cultural heritage, knowledge and traditions that are slowly diminishing due to the rapid tourism development.

In a sense, the Kubang Badak BioGeoTrail is a tool for achieving the sustainable development goals through geopark concept. However, the main challenge is not to develop the trail, but to make it succeed (Bolwerk 2018). Not only the scientific perspective on the possible economic and socio-cultural impacts that should be considered, but also the expectations of the local community. Apart from its scientific significance of the area, it is pertinent that the Kubang Badak BioGeoTrail will improve income and livelihoods and instil sense of pride and belonging, in order for the local community to have sense of ownership to conserve the area in a collaborative manner with relevant stakeholders. The importance of Kubang Badak BioGeoTrail in the economic, environmental, social and psychological aspects is pertinent towards ensuring its sustainability.

As observed from Fig. 8.1, the location of Kubang Badak is adjacent to Machinchang Cambrian Geoforest Park. Geoforest Park serves as a special conservation area within a permanent reserve forest with outstanding geological and biological resources where protection and wise utilisation of these resources are geared towards sustainable recreation promoting multidisciplinary research and enriching community awareness about the natural integration of various forest resources (Mohammad Ismail et al. 2004, 2005). Apart from Machinchang Cambrian, there is also Kilim Karst and Dayang Bunting Marble Geoforest Parks. Each of them is named after its geographic name and the most significant geoheritage features within the area (Leman et al. 2007a, b).

The following two sections (Sects. 8.4 and 8.5) will discuss first, steps taken to develop the Kubang Badak BioGeeoTrail through participatory approach, and second to examine learning outcomes from the development of Kubang Badak BioGeoTrail through the micro-experiences of local community, particularly from the members of Fishermen Co-operative of Sungai Kubang Badak (KPSP Kubang Badak) and Langkawi Tour Guides Association (LTGA).

8.4 Developing Kubang Badak BioGeoTrail

Initial discussion to propose Kubang Badak BioGeoTrail was held in 2015, where a group of four researchers from the Langkawi Research Centre (PPL), Universiti Kebangsaan Malaysia (UKM) together with representatives of KPSP Kubang Badak community presented to the Langkawi Development Authority (LADA) on the need to conserve and develop Kubang Badak in an innovative and sustainable manner. The island of Langkawi, at that time, had received approximately 2 million visitors annually, both international and domestic (LADA) (Langkawi District Local Plan 2030). Thus, it was a pertinent and timely discussion on Kubang Badak, as to ensure that sustainable alternative in future development of tourism attraction areas is shared together with all stakeholders, namely authorities, planners and policy-makers based on robust scientific research as well as local knowledge, values and concerns.

Subsequently, several discussions were held internally among researchers and local communities on developing the Kubang Badak BioGeoTrail framework in

2016. However, it was only in 2018 that fieldworks was initiated to identify potential sites for the trail, using materials gathered from discussions held earlier, supported by existing scientific literature on the geological, biological and cultural aspects of Kubang Badak, together with local knowledge shared by community of the area. The process of developing the trail was conducted using the five interrelated stages of project life cycle: needs assessment, conceptual design and feasibility, action planning, implementation and operation and maintenance. The involvement and participation of the local community in the first three stages (needs assessment, project design and planning) promote behavioural intentions to sustain the project. The success in implementing the first three stages in a way is to minimise challenges encountered at later stages (implementation and operation and maintenance), where at times the local community may have felt left out during the design and planning stages.

As a result of discussions and three fieldworks held in January, February and April 2018, 12 interesting sites accompanied by stories about the natural history of Kg. Kubang Badak, including cultural, geological and biological heritage were developed (Fig. 8.2). Identification of these 12 sites was conducted in a collaborative manner with local communities, local tour guides and four researchers from the Langkawi Research Centre (PPL), UKM as well as staffs from the Geopark Unit, Langkawi Development Authority (LADA). These sites as illustrated in Fig. 8.2 were developed with consideration to ensure integrated heritage conservation based on its significance as emphasised from three elements of cultural, geological and biological.

For instance, in site number 3, Pulau Kubang Badak highlights the biological, geological and cultural elements of the small island located in front of the fishermen jetty. The young island was formed by the deposition of fine silt alternating with mud, which is important in creating a young mangrove ecosystem, which, in turn, protects the jetty from strong wind and storms (Komoo et al. 2018). Meanwhile, at site 5, Kampung Siam, the emphasis on the cultural, biological and geological elements of the village once inhabited by the Southern Thai community. The charcoal industry was once a thriving activity carried out by the community until it ended in the 1980s, as it was no longer permitted by the Kedah Forestry Department (Komoo et al. 2018).

In addition, knowledge dissemination and documentation of these 12 sites were prepared in three forms: workshop for script writing related to each of the sites, workshop for local tour guide modules and preparation of booklet. A pilot study was also carried among undergraduate students of the UKM CITRA programme in May 2018. The objective was to test their understanding and effectiveness of the module prepared for local tour guides as part of the geotourism activities to be promoted in the area. All of this knowledge dissemination and documentation processes support the concept of Kubang Badak BioGeoTrail that promotes optimum and low impact tourism development. It was also during the workshops and booklet preparation that allows for researchers and local community to learn and understand each other capacity to define common ground and compromises (Fig. 8.3).

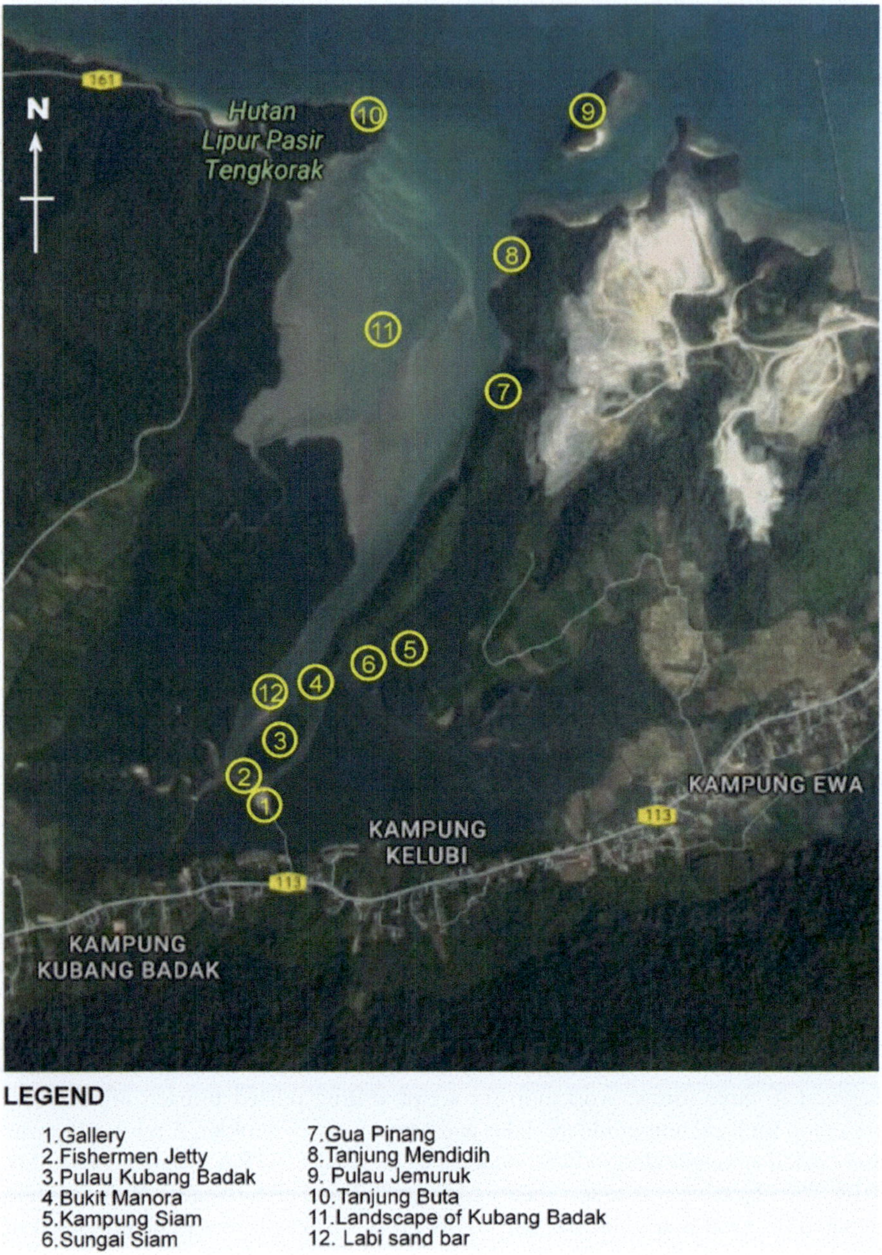

Fig. 8.2 The sites of Kubang Badak BioGeoTrail at Langkawi Global Geopark (source: Komoo et al. 2018)

Fig. 8.3 Local tour guide, En. Othman Ayeb, explaining the Kubang Badak BioGeoTrail to a group of conference participants (**a**) (photo by SA Halim), Pulau Kubang Badak (**b**) (photo by PPL UKM) and workshop held for Local Tour Guide on Kubang Badak BioGeoTrail Module (**c**) (photo by PPL UKM)

8.5 Lessons Learned

"What legacy are we leaving to future generations?" That was the question raised by the members of KPSP Kubang Badak when asked about the importance of conserving heritage. It was a question pertinent in opening opportunities for local community and researchers to developing the BioGeoTrail. In this chapter, micro-experiences of the local community are discussed to show the interconnectedness between human and natural resources as a source of well-being in developing the Kubang Badak BioGeoTrail. Overall, the interrelated stages of the project life cycle: needs assessment, conceptual design and feasibility, action planning, implementation and operation, and maintenance. The local community of Kubang Badak agreed that they contributed substantially to the BioGeoTrail research through transferring of tacit knowledge to the researchers.

Meanwhile, the transfer of scientific knowledge to the local community was demonstrated through changes in attitudes or behaviour, as described in Stage 1 and Stage 2 of the project life cycle. However, it is still a challenge to distinguish the effects of involvement in research and the effects of other knowledge sources or drivers of change in the community (Garnett et al. 2009). Stage 1 and Stage 2 are as follows:

Stage 1: Needs Assessment, Conceptual Design and Feasibility, Action Planning

In this stage, recognition given to local knowledge and scientific research acquired through the project increased the opportunities for intergenerational knowledge transfer, with older people passing skills to younger community members. There was also co-learning, with researchers learning and understanding more about not only their research topic but also about local context. Some of the voices from the community related to Stage 1 are shared:

> We are able to share with people about our knowledge of Kubang Badak through this BioGeoTrail as a platform

> Making the folklore of Kubang Badak alive, such as in Bukit Menora, whilst conserving the hill that looks almost like a cone (mogote, a geological term for an isolated hill with steep slopes formed by either limestone, marble or dolomite)

> Potential to improve our livelihoods through innovative activities, encourage community to work together in finding ways to generate income

Local tales of Kubang Badak are unique to the environment and related to the socio-cultural origin of the community. Place names and place making are useful for enhancing local community identity and sense of belonging to the heritage conservation efforts (Abdul Aziz and Liu 2011). Identification of the 12 sites in the trail was an opportunity for locals to get to know the historical aspects of their local population. The interpretation of landscape from a scientific perspective together with cultural knowledge has revealed an intrinsic link between geology, biology and its people, not only for conservation purpose but also to boost local socio-economic development (Unjah and Halim 2018).

Generally, responses from the local community in developing the Kubang Badak BioGeo trail demonstrated their commitment to volunteer their willingness to invest personal resources, such as time and energy in working together with researchers. This finding concurs to Olson (1971) who postulated that leadership amongst local community emerges because they value the collective good more and are willing to incur costs to participate. Even so, Kubang Badak BioGeoTrail is still at an early stage of commercial development, more research is needed to understand the socio-economic and social-cultural effects of the trail development, particularly in addressing poverty alleviation in rural communities.

Stage 2: Implementation and Operation, and Maintenance

The development of Kubang Badak BioGeo trail, serves as evidence of community-based management or co-management of natural resources could be done effectively, whilst enhancing social capital and increasing incomes of households (Garnett et al. 2009; Popa et al. 2017; Abdul Halim and Ishak 2017).

In Stage 2, technical skill development and capacity-building process are vital to be given as part of the training to local community to sustain the trail effectively. The voices shared from the younger age group of the Kubang Badak community as illustrated below highlight the many aspirations expressed by local community and the need for a more strategic approach and implementation of the trail:

> More efforts to equip our people with knowledge and skills, in order to take care of our trail, both from internal and external aspects

> Kubang Badak is not like others, we want to make a dream come true here

> Not only focus on the BioGeoTrail per se but also linking it with Kubang Badak and its surrounding areas for inclusive and benefits for all

Training is not only on conservation but also other related topics such as finance management, how to start micro-business, communication, leadership, health and safety. Furthermore, the development of Kubang Badak BioGeoTrail allows for greater monitoring of environmental changes in the area with local authorities and expanding how local community understands the impacts and consequences in conservation with development, particularly for heritage conservation.

Developing the Kubang Badak BioGeoTrail in a collaborative and participatory approach, continuous efforts are still needed to investigate the environmental and social trajectories through action research fusing local knowledge and scientific research. Ensuring all stakeholders commitment and momentum are critical to maintaining the governance and management of the trail, particularly with locals and institutional support, not only financially but also in terms of moral support and obligation to ensure the success of the BioGeoTrail development.

8.6 Conclusion

The micro-experiences shared from the Kubang Badak BioGeoTrail suggest that appreciation towards natural resource management, and development, is as much about people as it is about the natural environments, and that modes of knowledge transfer need to acknowledge social context. Therefore, it is with great hope that the development of this Kubang Badak BioGeoTrail serves as evidence and motivation to encourage greater public engagement in conservation practice together with experts, authorities and relevant stakeholders in achieving sustainability aspirations for the sake of present and future generations.

Acknowledgements We would like to acknowledge the support given from the grant code S170531 entitled "A Study to Developing BioGeoTrail Kubang Badak, Langkawi, Kedah". Our utmost appreciation to all local communities of Kg. Kubang Badak whom have generously facilitated the research process, especially to Mr. Othman Ayeb, Puan Indra Rahayu, Mr. Sabri Musa and Mr. En. Rusasmizal Mohd Ghazali for their tireless efforts in promoting heritage conservation for it to be cherished by the present and future generations.

References

Abdul Aziz R, Liu OP (2011) Placemaking, place names, and local myths and legends. Plan Malays. https://doi.org/10.21837/pm.v11.95

Abdul Halim S, Ishak NA (2017) Examining community engagement in heritage conservation through geopark experiences from the Asia Pacific region. Kajian Malaysia 35(Suppl. 1):11–38

Adom D (2018) The human impact and the aquatic biodiversity of lake Bosomtwe: rennaisance of the cultural traditions of Abono (Ghana)? Trans Rev Syst Ecol Res 20(1):87–110

Adom D (2019) The place and voice of local people, culture, and traditions: a catalyst for ecotourism development in rural communities in Ghana. Sci Afr 6. https://doi.org/10.1016/j.sciaf.2019.e00184

Arnstein A (1969) A ladder of citizen participation. J Am Inst Plann 35(4):216–224

Bennett NJ, Roth R, Klain SC, Chan K, Christie P, Clark DA, Cullman G, Curran D, Durbin TJ, Epstein G, Greenberg A, Nelson MP, Sandlos J, Stedman R, Teel TL, Thomas R, Veríssimo D, Wyborn C (2017) Conservation social science: understanding and integrating human dimensions to improve conservation. Biol Conserv 205:93–108

Bolwerk M (2018) Local socio-economic and socio-cultural effects of geotourism: an analysis of the BioGeoTrail in Kubang Badak, Malaysia. In: Unpublished BA Thesis 2017/2018. University of Groningen, Groningen

Burke BJ, Welch-Devine M, Sourdrill A (2020) Understanding microexperiences of climate change: how climate ethnography informs collaboration, adaptation, and effective responses. In: Welch-Devine M et al (eds) Changing climate, Changing worlds, Ethnobiology. Springer, Cham. https://doi.org/10.1007/978-3-030-37312-2_1

Chitty G (ed) (2017) Heritage conservation and communities: engagement, participation and capacity-building. Routledge, Abingdon

Conrad E, Cassar LF, Jones M, Eiter S, Izaovicova Z, Barankova Z, Christie M, Fazey I (2011) Rhetoric and reporting of public participation in landscape policy. J Environ Pol Plan 13(1): 23–47

Farsani NT, Coelho C, Costa C (2011) Geotourism and geoparks as novel strategies for socioeconomic development in rural areas. Int J Tour Res 13:68–81

Fuller D, Caldicott J, Cairncross G, Wilde S (2007) Poverty, indigenous culture and ecotourism in remote Australia. Development 50(2):141–148

Garnett ST, Crowley GM, Hunter-Xenie H, Kozanayi W, Sithole B, Palmer C, Southgate R, Zander KK (2009) Transformative knowledge transfer through empowering and paying community researchers. Biotropica 41(5):571–577

Global Geopark Network (GGN) (2010) Guidelines and criteria for National Geoparks Seeking UNESCO's Assistance to Join the Global Geoparks Network

Gonzalez-Tejada C, Du Y, Read M, Girault Y (2017) From nature conservation to geotourism development: examining ambivalent attitudes towards UNESCO directives with the Global Geopark Network. Int J Geoheritage 5(2):1–20

Halim SA, Komoo I, Salleh H, Omar M (2011) The geopark as a potential tool for alleviating community marginality. Shima Int J Res Island Cult 5:94–113

Hashim HS, Aziz S, Abdul Aziz R (2011) Conservation with development: showcasing Langkawi Geopark. Planning Malaysia

Hes D (2017) Impact of community engagement on sustainability outcomes–there is no sustainability without community engagement. https://msd.unimelb.edu.au/data/assets/pdf_file/0006/3027795/Next_Gen_Expert_series_D_Hes-1c4ud7u.pdf (accessed 3 November 2019)
Ilmonen K (2015) The role of culture in regional development work—changes and tensions. In: Go FM, Lemmetyinen A, Hakala U (eds) Harnessing place branding through cultural entrepreneurship. Palgrave Macmillan, London, pp 79–95
Komoo I, Patzak M (2008) Global Geoparks Network: an integrated approach for heritage conservation and sustainable use. In: Leman MS, Reedman A, Chen SP (eds) Southeast and East Asia geoheritage conservation. LESTARI UKM, Bangi, pp 1–13
Komoo I, Ahmad N, Unjah T, Abdul Halim S (2018) Kubang Badak heritage trail. LESTARI, UKM, Bangi
Langkawi District Local Plan (Rancangan Tempatan Daerah Langkawi). 2030
Leman MS, Abdul Ghani K, Komoo I, Ahmad N (2007a) Langkawi Geopark. LESTARI UKM, Bangi
Leman MS, Komoo I, Mohamed KR, Ali CA, Unjah T (2007b) Geopark as an answer to geoheritage conservation in Malaysia—The Langkawi Geopark case study. Geological Society of Malaysia, Bulletin 53:95–102
Li N (2010) Preserving urban landscapes as public history: the Chinese context. Public Hist 32:51–61
Mahjabeen Z, Shresha KK, Dee JA (2009) Rethinking community participation in urban planning: the role of disadvantaged groups in Sydney metropolitan strategy. Australas J Reg Nal Stud 15(1):45–63
Majlis Perbandaran Langkawi Bandaraya Pelancongan (2002) Draf Rancangan Tempatan Daerah Langkawi 2001-2015
Mohamad Ismail S, Komoo I, Leman MS, Mohamed KR, Ali CA, Ahmad N, Latiff A (eds) (2005) Geoforest parks–Hanging Garden of Langkawi. Forestry Dept Peninsular Malaysia and Lestari UKM, Kuala Lumpur
Mohammad Ismail S, Komoo I, Leman MS (2004) Geo-forest park: an innovative approach towards geological heritage conservation within permanent reserved forests of Malaysia. In: Leman MS, Komoo I (eds) Geological heritage of Malaysia—theoretical framework and geoheritage assessment. LESTARI UKM, Bangi, pp 243–250
Nared J, Bole D (2020) Participatory research on heritage- and culture-based development: a perspective from south-east Europe. In: Nared J, Bole D (eds) Participatory research and planning in practice. The Urban Book Series. Springer, Cham. https://doi.org/10.1007/978-3-030-28014-7_7
Nasser N (2003) Planning for urban heritage places: reconciling conservation, tourism, and sustainable development. J Plan Lit 17(4):467–479. https://doi.org/10.1177/088541220301700400l
Olson M (1971) The logic of collective action: public goods and the theory of group. Harvard University Press, Cambridge
Ostrom E (1990) Governing the Commons: the evolution of Institutions for collective action. Cambridge University Press, New York
Popa R, Popa D, Andrasanu A (2017) The SEA and Big-S models for managing geosites as resources for local communities in the context of rural geoparks. Geoheritage 7:175–186
Pretty JN (1995) Participatory learning for sustainable agriculture. World Dev 23(8):1247–1263
Rancangan Tempatan Daerah Langkawi (RTDL) (Langkawi District Local Plan). 2030. 29 Dec 2016 (Gazetted)
Reed J, Barlow J, Carmenta R, Vianen, van Vianen, J., Sunderland, T. (2019) Engaging multiple stakeholders to reconcile climate, conservation and development objectives in tropical landscapes. Biol Conserv 238:108229
Ronchi S, Arcidiacono A, Pogliani L (2020) Integrating green infrastructure into spatial planning regulations to improve the performance of urban ecosystems. Insights from an Italian case study. Sustain Cities Soc 53:101907., ISSN 2210-6707. https://doi.org/10.1016/j.scs.2019.101907

Sacco P, Ferilli G, Blessi GT (2013) Understanding culture-led development: a critique of alternative theoretical explanations. Urban Studies 51:2806–2821. https://doi.org/10.1177/0042098013512876

Shea M (2019) Participatory methods and community-engaged practices for collecting, presenting, and representing cultural memory. In: Drozdzweski D, Birdsall C (eds) Doing memory research. Springer, Singapore. https://doi.org/10.1007/978-981-13-1411-7_9

Tan H, Bouchaud J (2015) Langkawi style. The Lemongrass, Penang

United Nations. Economic and Social Commission for Asia and the Pacific (2003) Poverty alleviation through sustainable tourism development. United Nations, New York

Unjah T, Halim AS (2018) Connecting legend and science through geomythology: case of Langkawi UNESCO Global Geopark. Kajian Malays 35(Suppl. 1):77–89

Walter RK, Hamilton RJ (2014) A cultural landscape approach to community-based conservation in Solomon Islands. Ecol Soc 19:41. https://doi.org/10.5751/ES-06646-190441

Wetlands International Malaysia (2005) Rapid ecological assessment of the Kubang Badak mangroves and associated wetland ecosystems, Pulau Langkawi, Malaysia (Project No. MYS 441/00)

Zhang Y, Xiao X, Cao R, Zheng C, Guo Y, Gong W, Wei X (2020) How important is community participation to eco-environmental conservation in protected areas? From the perspective of predicting locals' pro-environmental behaviours. Sci Total Environ 739:139889. https://doi.org/10.1016/j.scitotenv.2020.139889

The Dynamics of Environmental Change Pose Challenges to Preserving the Biocultural Landscape in Indonesia

9

Zulfaidah Penata Gama and Elita Agustina

Abstract

Environmental changes can be affected by climate change, global warming, or disasters. They typically can lead to changes in the landscape, the ecosystem, and their natural habitat. Global warming is the process of an excessive increase in the Earth's temperature. The cause is, of course, excessive greenhouse gases. Climate change is an extreme reaction to weather phenomena that negatively affect agricultural resources, water resources, human health, depletion of the ozone layer, vegetation, and soil, causing double the concentration of carbon dioxide in ecosystems. The dynamics of environmental change also occur due to industrial development in urban areas. There are also many changes in rural areas in Indonesia because many people are building houses in the village area. Villages considered having pollution-free air and a healthy environment. The environmental changes that occur also have an impact on biocultural changes. Therefore, the effort of maintaining balance and biocultural preservation is aimed at maintaining local wisdom in each region. Therefore, environmental change dynamics provide an opportunity to maintain the biocultural landscape while maintaining local wisdom in each region. Biocultural landscapes are interrelated holistic systems that have been formed by human management for a long time. In Indonesia several provinces have experienced disasters such as tsunami. Tsunami impacts

Z. P. Gama (✉)
Department of Biology, Faculty of Mathematics and Natural Sciences, Universitas Brawijaya, Malang, East Java, Indonesia
e-mail: gama@ub.ac.id

E. Agustina
Department of Biology Education, Faculty of Education and Teacher Training, Universitas Islam Negeri Ar-Raniry, Banda Aceh, Aceh, Indonesia
e-mail: elita.agustina@ar-raniry.ac.id

S. A. Abdullah et al. (eds.), *Conserving Biocultural Landscapes in Malaysia and Indonesia for Sustainable Development*,
https://doi.org/10.1007/978-981-16-7243-9_9

sudden changes in the landscape, which will inevitably cause the population around the disaster site to adapt to survive. Likewise, animals and plants affected by disasters experience very significant changes. In a healthy biocultural landscape, pets are an extension of the human community, an essential part of the growing and ever-strengthening relationship between humans and the natural world. The knowledge and wisdom of indigenous groups have developed over the centuries and are interdependent with their natural environment. These are sometimes referred to as “indigenous,” “local,” or “cultural knowledge,” traditional knowledge, or local wisdom and are an understanding of nature, such as the nature of plants, animal behavior, ecosystem balance, and food and medicine. Medical problems are also sometimes related to ethnic issues in an area. Traditional knowledge is essential for the resilience of a biocultural system. Efforts to revitalize, share, respect, and implement these policies are critical for protecting biodiversity, preserving cultural heritage, maintaining global ecological health, and mitigating climate change. This chapter is aimed at describing the importance of local wisdom actions to restore post-tsunami conditions in an area in Aceh that affects the biocultural landscape in that location.

Keywords

Aceh · Biocultural landscape · Disaster adaptation · Local wisdom · Tsunami

9.1 Introduction

Global warming is a symptom of climate change. At this time, global warming and climate change are not new issues heard by the Indonesian people. Global warming occurs when the global average temperature or the temperature of the Earth’s surface increases. Global warming is a rapid increase in the Earth’s average surface temperature over the last few years, mainly due to the greenhouse gases released when humans burn fossil fuels (Bhandari 2018). The gradual increase in the Earth’s atmosphere, commonly associated with the greenhouse effect, is caused by increasing carbon dioxide, chlorofluorocarbon (CFC), and other pollutants. Global warming is the most severe environmental problem and can cause temperature rise, sea-level rise, climate change, and ecological change that profoundly affect human existence.

Indonesia is prone to disasters primarily related to climate change (Measey 2010). Other causes such as forest and land fires, landslides, storms, and droughts have destroyed infrastructure and damaged forest and coastal ecosystems, resulting in loss of life and property ecosystem services and livelihoods. Most of Indonesia’s greenhouse gas emissions come from changes in land use, forests, energy, agriculture, waste, and industrial processes.

Industrialization is a process of socio-economic change that changes the system’s livelihood of an agrarian society to be an industrial society. Automation can be interpreted as a state where association focuses on one’s economy, covering increasingly diverse jobs (specialization), increasing salaries, and income high.

Industrialization is part of modernization, where social change and economic development are closely related to technological innovation. The industrial revolution had a positive impact on the world economy because the industrial revolution made the price of goods cheaper and the work easier. After all, machines assist in it. But the influence of the industrial revolution on climate change on Earth is negative. Because of to the increasing number of factories that produce waste and smoke, this can result in higher levels of environmental pollution. This pollution in general can have an impact such as acid rain which can damage forests and agricultural land.

Indonesia's position on the Pacific Ring of Fire (an area with much tectonic activity) causes the country to face many risks in the form of natural hazards such as volcanic eruptions, earthquakes, floods, and tsunamis. Underwater earthquakes or volcanic eruptions in the ocean can cause tsunami waves that are harmful to people and the sea environment (Mutaqin et al. 2019). In 2004, most of the world was hit by the Indian Ocean earthquake and tsunami. This incident killed more than 167,000 people in Indonesia (mainly Aceh) and displaced more than half a million people as thousands of their homes were destroyed. Although massive tsunamis like the 2004 tsunami are rare, the Sumatra region is often shocked by an offshore earthquake that can trigger a tsunami. The tsunami wave destroyed everything in its path, such as boats, buildings, bridges, cars, trees, telephone lines, and power lines.

The tsunami's caused damage to human life and hurt insects, animals, plants, and natural resources. The tsunami has also changed landscapes. Tsunamis can destroy trees, plants, and animal habitats such as bird nests. Land animals are killed because the tsunami waves drown them, and marine animals die of pollution when the harmful chemicals are washed into the sea, poisoning marine life. The impact of the tsunami on the environment is related to the landscape, animal life, and human-made environmental aspects. Solid waste and debris from disasters are the most critical environmental problems facing a tsunami-stricken country.

Before the tsunami, every area had biodiversity, unique biocultural values, and a population with diverse traditional ecological knowledge that had been preserved sustainably for decades. Humans have used the surrounding landscape and living organisms as their source of life. Besides, humans have developed a new variety through cultivation. Biodiversity forms the foundation that facilitates the formation of cultural diversity. At the time of the tsunami, each affected area experiences a change in landscape and loss of inherited knowledge of biocultural values. Efforts to maintain biocultural diversity are very urgent because the threat of loss of indigenous cultures continues over time (Martinez-Reyes 2012).

Post-tsunami development of buildings, settlements, roads, and bridges in Aceh has some impact on changes in the biodiversity and socio-culture of the community. Rapid changes in the ecosystem of an area indicate an ecological imbalance between humans and nature. Efforts have been made to restore and maintain a balanced level of biodiversity and landscapes can be handled at the level of cultural diversity or local wisdom. Many studies have demonstrated that the human adaptation process is rooted in biological resources and the traditional knowledge that is created. Biodiversity and cultural diversity that exist in a particular area can be said to be natural resources that form the basis for determining the cultural characteristics of the community. The knowledge held by local people can also be considered a valuable resource. The ability of humans to respond to changes in the global environment can be predicted through the development and dissemination of customs. Thus, it is true that humans and nature coexist, interact, and depend on one another in an ecosystem. Therefore, it is very important to understand the connection between people and nature, which is affected by impacts such as that of the tsunami (Hong 2011).

The idea of relating to biocultural systems arose from landscape geographers' and ecologists' work on more mechanical socio-ecological systems and more human-centered cultural landscapes. This concept also grew politically through the Convention Concerning the Protection of the World Cultural and Natural Heritage (UNESCO 2017) and the impetus that arose from that international treaty. The World Heritage Convention helps focus attention by relating designations that were previously separate from natural and cultural sites through framing landscapes in biocultural contexts. Different cultures and societies viewed and valued biodiversity differently because of their distinct heritage and experiences. Much of the discussion about the complicated relationship between the conservation of biodiversity and cultural diversity has been centered on the argument that cultural diversity can underpin multiple natural resource use and protection practices. Ecosystem services can also be seen through the prism of nature's contributions to people (Díaz et al. 2018).

Biocultural landscapes are formed by dynamic processes developed by nature and humans. They connect with humans through various interactions, creating complex biocultural systems full of functions, services, and values. The intensive and rapid modification of the biosphere in the context of global change by humans is one of the tasks of landscape ecology, namely determining the method of analyzing the interaction of humans and modified ecosystems, taking into account the consequences of the sociocultural and human economy (tradition, local wisdom, welfare) and ecosystem (integrity, resilience), and providing indicators for these interactions and effects.

Local wisdom plays a role in managing natural resources and the environment. However, local wisdom also has the face of several challenges, such as population growth, modern technology and culture, enormous costs, poverty, and inequality. The prospects for local wisdom in the future depend heavily on community knowledge, technological innovation, market demand, utilization and preservation of biodiversity in the environment, government policies that are directly related to the management of natural resources and the environment, and the role of the community. The role of the community in maintaining and preserving local wisdom values must continue to the next generations, for example, the experience of the community in Aceh Jaya district, Aceh Province. The community applies local wisdom values after the 2004 Aceh tsunami disaster. The efforts made by the Aceh Jaya community to prevent a disaster were to prohibit the community from cutting down trees around rivers, that is, direct community monitoring of forest use. The community independently protects and conserves the forest by planting trees. The government as the policymaker needs to ensure that the values of disaster-based local wisdom will continue to exist amid society through various technological and educational innovations (Erianjoni 2016). This chapter aims to describe the importance of local wisdom actions to restore post-tsunami conditions in an area in Aceh that affects landscape bioculture in that location.

9.2 Case Study and its Site: Tsunami Disaster in Aceh

Based on historical records, Aceh has experienced several tsunamis, namely in 1837, 1861, 1885, 1886, 1907, 1922, 1949, 1964, 1967, 2004, 2010, and 2012 (BMKG 2019). Only the tsunami that occurred in December 2004 has the most complete data related to the impact of the tsunami. The tsunami in 2004 resulted in 130,000 deaths, 37,000 people missing, and 500,000 houses damaged, and it hit all districts/cities in the coastal areas of Aceh, especially the west coast (BRR and Patners 2006). Based on tsunamis that have occurred several times in Aceh Province, it can be categorized as a disaster that has a permanent tendency (BPBA 2019).

The tsunami disaster that occurred on December 26, 2004, has destroyed the coastal area of Aceh Province. It caused the loss of human lives and damaged infrastructure, settlements, public facilities, and coastal ecosystems. The damage to the coastal ecosystems includes the intrusion of seawater and silt to land, the destruction of coral reefs, the uprooting of some coastal vegetation, changing coastline, and wetland morphology.

The tsunami in 2004 was one of the most devastating tsunamis that resulted from a massive earthquake. The earthquake that triggered this tsunami was between 9.1 and 9.3 Mw (Poisson et al. 2011). Aceh Province is one of the provinces most affected by the tsunami in 2004. The worst effect of the tsunami was reported in Banda Aceh, Aceh Jaya, Aceh Besar, and Aceh Barat. This tsunami also hit several other areas to the east of Aceh, such as Pidie, Bireuen, and Lhokseumawe. The east coast of Aceh has been affected by a tsunami that is smaller than the area in Banda Aceh and the West–South coast of Aceh (Meilianda et al. 2019).

9.3 The Effect of the Tsunami Disaster on the Aceh Coastal Ecosystem

A tsunami as high as 10–15 m traveling at a speed of 40 km/h causes damage and destroyed the coast of Aceh, namely along the West Coast of Aceh, the 21 districts/cities in Aceh Province, 15 of which were hit by the earthquake and tsunami. The total area affected by the earthquake and tsunami reached 649,582 ha, including 61,816 ha of agricultural land, 11,609 ha of aquaculture ponds, and 32,004 ha of mangrove forests and coastal vegetation (UNEP 2007a).

The earthquake and tsunami in Aceh had a major and comprehensive impact on the people of Aceh at various levels. We can see the impact of the destruction of residential infrastructure, community services, and the natural environment in particular coastal areas. The tsunami's effect on the coastal and marine regions of Aceh Province caused pollution of the sea (material flakes), the groundwater, and a change in the coastline. Mangrove forests that function as protection for settlements from waves and winds were severely affected by tsunami waves. Mangrove forest, which is nature protection, also serves as a spawning ground, nursery ground, and feeding ground for various marine life, including fish. Another natural protection that has been damaged and lost is coral reefs that function as shelters and fish spawning grounds. This condition has an impact on reducing fish resources and coastal species (potential for biodiversity). Damage and loss of wetland ecosystems and artificial ecosystems have a significant effect on people's economic activities (Rotty 2018).

Apart from being caused by the tsunami, damage to the coastal areas was also caused by earthquakes that had changed the landscape along the West Coast of Aceh and Simeulue Island. Two mechanisms caused damage to the coastal ecosystem due to the tsunami, namely tsunami energy, which hit the coast and destroyed mangrove forests, pine (sea pine) stands, coconut groves, and various other vegetations. Even in some coastal areas that were severely affected by the tsunami, the waves' force also caused the mangrove trees to be uprooted and did not leave any vegetation around them. Another mechanism that causes coastal vegetation damage is standing seawater that remains on land after the tsunami. Stagnant seawater that has high salinity affects the growth and development of coastal plants, which become stressed and dry and then die (Wibisono and Suryadiputra 2006).

Damage to coastal ecosystems such as mangroves and coral reefs has caused the loss of protection of lowland coastal areas. The existence of coastal ecosystems is essential in protecting the coast from extreme waves caused by wind, including cyclone storm surges. However, information regarding coastal ecosystems' role in protecting the land from tsunami waves is still minimal. The coastal ecosystem can be an essential predictor variable for the tsunami hazard. Other coastal ecosystem components that can act as a buffer against the tsunami hazard are coral reefs, seagrass beds, mangroves, and other vegetation that protects against disasters (Cochard et al. 2008).

9.3.1 Post-tsunami Coastal Vegetation in Aceh

Aceh Province has two types of vegetation: upland tropical rainforest vegetation and lowland tropical rainforest vegetation. The lowland tropical rainforest has a vital role in protecting coastal areas. The types of vegetation are coastal vegetation that is experiencing elevation or formation of escape, coastal vegetation that is experiencing erosion, and rocky beach vegetation (Suryawan 2007).

The tsunami disaster caused the coastline to be severely damaged, and almost all vegetation was destroyed and lost. Damaged coastal vegetation included mangrove vegetation, coastal vegetation, and lowland tropical rain vegetation. The vegetation of the damaged coastal areas has experienced changes (succession), by the emergence of pioneer plant species such as herbs, shrubs, and saplings. However, the coastal province of Aceh has experienced abrasion from year to year. Efforts to mitigate coastal erosion are carried out by planting potential plants in coastal areas and preserving coastal vegetation. The effort was initiated by the government and

A B C

Fig. 9.1 Species that can currently be found on the Lhoknga Aceh coast (**A**. *Ipomoea* spp., **B**. *Hibiscus tiliaceus,* and **C**. *Casuarina equisetifolia*)

private in several locations in Aceh through a program of planting by community groups. The kinds of species that are used for coastal rehabilitation plants are *Cocos nucifera*, *Casuarina equisetifolia*, *Terminalia cattapa*, *Cerbera manghas*, *Hibiscus tiliaceus*, *Azadirachta indica*, *Callophylum inophyllum*, *Jatropha curcas*, and *Pandanus tectorius* (UNEP 2007b).

Aceh's East Coastline is 761 km long and has a muddy beach type. Before the tsunami, this beach was covered by a large area of mangrove forest. Meanwhile, the west coastline of Aceh is 706 km long and is dominated by sandy beaches covered with coconuts, sea cypresses, and other types of coastal vegetation. Aceh's West Coast stretches from Banda Aceh City to Nagan Raya. Based on the local community's information, the dominant species before the tsunami along the West Coast of Aceh were *Hibiscus tiliaceus*, *Pterospermum diversifolium*, *Casuarina equisetifolia*, *Ficus septica*, *Pongamia pinnata*, *Cerbera manghas*, *Timonus compressi caulis, Barringtonia asiatica*, and *Ipomoea* spp. Before the tsunami, *Ipomoea* spp. grew on the back of a sandy beach, but it was found growing far inland after the tsunami. *Ipomoea* spp. can also serve as a biological indicator that the area is suitable for planting *Barringtonia asiatica, Callophyllum innophyllum, Casuarina equisetifolia, Cerbera manghas, Pongamia pinnata*, *Terminalia catappa,* and *Hibiscus tiliaceus*. The community plantations located around the coast before the tsunami was coconut (*Cocos nucifera*) and rubber (*Hevea brasiliensis*) plantations (Wibisono and Suryadiputra 2006). Species that can currently be found on the Lhoknga Aceh coast can be seen in Fig. 9.1.

9.4 Post-tsunami Mangrove Damage

Mangrove forests are vegetation that is generally found on muddy tropical beaches such as the coastal areas of Aceh. Aceh Province has mangrove forests that are scattered along the east coast and West Coast of Aceh. The east coast of mangrove forests can be found in Aceh Tamiang, East Aceh, North Aceh, and Bireun. Meanwhile, on the west coast, you can find mangrove forests in Aceh Jaya, West Aceh, and Aceh Singkil. Pulau Simeulu and Pulau Banyak are islands with mangrove forests that were also affected by the tsunami (Wibisono and Suryadiputra 2006).

Several types of plants found in Aceh's mangrove forests before the tsunami hit were *Avicennia marina*, *A. officinalis*, *A. alba*, *A. lannata*, *Bruguiera gymnorrhiza*, *B. parviflora*, *Ceriops tagal*, *C. decandra*, *Lumnitzera littorea*, *L. racemosa*, *Rhizophora mucronata*, *R. apiculata*, *R. stylosa*, *Scyphiphora hydrophyllacea*, *Sonneratia alba*, *S. caseolaris*, *Excoecaria agallocha*, *Aegiceras cornoculatum*, *Xylocarpus rumphii*, and *X. granatum* (Wibisono and Suryadiputra 2006). Because of the zoning in the mangrove forest, the vegetation that dominates is different. The zone closest to the sea is dominated by *Avicennia* spp., which can tolerate high salinity levels. By contrast, the middle area (mesozone) is dominated by *Ceriops* spp., *Lumnitzera* spp., *Scyphiphora hydrophyllacea*, *Bruguiera* spp., and *Rhizophora* spp. In the back zone, further into the drier soil conditions, the dominant species are *Aegiceras* and *Xylocarpus* that grow well (Noor et al. 2006).

The tsunami in 2004 destroyed the ecosystem on almost the entire coastline of Aceh. The strength of the tsunami waves destroyed tens of thousands of hectares of mangrove forests. Before the tsunami disaster, Aceh's mangrove forests had undergone extensive degradation. The leading cause is the development of shrimp ponds, oil palm plantations, and mangroves' felling for charcoal production (Wibisono and Suryadiputra 2006). Mangrove forest degradation in Aceh got worse after the tsunami. Mangrove forests in Aceh, which have 350,000 ha, suffered damage of around 50.3% or 174,590 ha due to the tsunami (Rotty 2018). Various species that grow in the middle of the mangrove forest can be seen in Fig. 9.2.

A B C D

Fig. 9.2 Various species that grow in the middle of the mangrove forest (Source: Wibisono and Suryadiputra 2006) (**A**. *Scyphiphora hydrophyllacea*, **B**. *Ceriops* spp., **C**. *Bruguiera gymnorrhiza*, and **D**. *Lumnitzera littorea*)

9.5 Preservation of the Biocultural Landscape in Aceh

The effort to rehabilitate and reconstruct tsunami-affected areas is a difficult activity to do. This activity requires a long time and sustainable effort. Implementing the rehabilitation and reconstruction program for a tsunami disaster area requires studies from various aspects of local community life, such as social, economic, cultural, and environmental aspects. Efforts to achieve the expected goals and targets require commitment from various parties, both government and non-government. The aim of the rehabilitation and reconstruction of tsunami-affected areas is to protect coastal ecosystems' uniqueness and ecological function and improve community livelihoods. It is hoped that the recovery of the coast's natural conditions can restore the livelihoods of the community such as fisheries, livestock, ecotourism, and other new resources while maintaining and preserving the environment based on Acehnese local wisdom.

9.5.1 Rehabilitation of Mangroves and Surrounding Animal Habitats

Mangrove replanting takes place along the east to the West Coast of Aceh. The approach taken involves the community in rehabilitating the land from the planning, preparation, planting, and maintenance stages of seedlings. The goal of post-tsunami coastal community empowerment is to immediately restore the coastal ecosystem's ecological condition and increase alternative livelihoods for people who depend on coastal natural resources. One of the government's efforts to improve the standard of living of the community is the provision of environmentally friendly grants, namely integrating mangrove ecosystem restoration with community-based small-scale economic activities (Rotty 2018). In 2006, 30 million mangrove seedlings have been planted in areas affected by the tsunami. The government rehabilitated a mangrove forest covering an area of 26,000 ha in 2010. Regeneration that has been carried out indicates that natural recovery and succession of mangrove forests is slower and more complicated than regeneration and succession of coastal forest vegetation. The habitat has been destroyed and drastically changed, making it unsuitable for mangrove forests. Habitat changes affect mangrove forests, such as physical and chemical factors, growth media or substrate, loss of mother trees, and the seawater ebb and flow (Wibisono and Suryadiputra 2006).

Apart from mangrove restoration, an environment-based shrimp cultivation program is also being carried out. Aceh is one of the major shrimp commodity-producing provinces in Indonesia. However, in 2000, this commodity had experienced a decline and continued to decline after the earthquake and tsunami in Aceh. Efforts to restore the productivity of the shrimp commodity in Aceh are rehabilitation of the ponds. The community is assisted in implementing environment-friendly shrimp farming to reduce pond rehabilitation activities' negative impact on environmental quality (Rotty 2018).

9.5.2 Cultivation of Coastal Forest Types

In 2008, coastal forest rehabilitation had reached 1000 ha after the tsunami with the planting of 1.54 million plants. This rehabilitation program involves the community through the coastal plant program in Aceh. This program, which involves the community, can rehabilitate coastal ecosystems, develop community livelihoods, and make village regulations that can support coastal ecosystem rehabilitation efforts and environmental education campaigns.

One of the districts most severely affected by the earthquake and tsunami in Aceh Jaya district is located on the West Coast of Aceh. The process of rehabilitating coastal forests in Aceh Jaya district is carried out by planting 371,750 plants consisting of 19 plant species. The tsunami has caused the retreat of the coastline up to 500 m inland, destroyed coastal forest vegetation and swamps, and turned rice fields into swamps. This area now forms a stretch of mangrove swamp where people catch shrimp and fish. The coastal forest rehabilitation program in Aceh Jaya district has obstacles in its implementation. This condition was caused by the activity of excavating sand for building materials after the tsunami that causes rising sea levels about 3 m from the previous limit during the tide. This rising sea level causes some of the newly planted coastal plants to die. However, in some places, coastal forest vegetation has naturally started to grow and develop.

Restoration of coastal forest resources improves the quality of the surrounding environment. This rehabilitation is also likely to bring back turtles to lay their eggs along the coast of Aceh Jaya Regency, as they did before the earthquake and tsunami disaster. The plant species that have been planted by the community on the coast of Aceh Jaya Regency in 2007 are from groups of mangroves, coastal plants, and forest plants. The mangrove plant groups planted by the community are large mangroves (*Rhizophora mucronata*), red mangroves (*R. apiculata*), Yellow Mangrove (*Ceriops tagal*), and api-api (*Avicenia*). The types of coastal plants that have been planted are sea cypress (Casuarina *equisetifolia*), mastwood (*Calophyllum innophyllum*), coconut (*Cocos nucifera*), sea hibiscus (*Hibiscus tiliaceus*), Indian Almond (*Terminalia cattapa*), and Indian Ash Tree (*Lannea coromandalica*). Meanwhile, the residents' yard plants are Balinese orange, chocolate, durian, mango, rambutan, starfruit, tamarind, and jambu keling (WIIP and WWF Indonesia 2008).

9.6 The Importance of Local Wisdom in Aceh

Disaster management based on local wisdom is an effort to optimize the natural potential and local cultural values of the community to facilitate the rehabilitation process after natural disasters such as the earthquake and tsunami in 2004. Aceh, as one of the disaster-prone areas in Indonesia (BPBA 2019), has several values of local wisdom, a combination of religion, local culture, and natural potential. Among the local wisdom implemented at the reconstruction and rehabilitation stage after the tsunami disaster in Aceh is the construction of a housing structure that follows the Aceh traditional house (*Rumoh Aceh*), replanting coastal forests in Aceh (*Uteun*

Fig. 9.3 Construction of houses following the Aceh traditional houses (*Rumoh Aceh*) (**A**. before tsunami, **B**. after tsunami)

Aceh), cultivation of marine biota, and coastal and marine environmental management by *Panglima laot* based on the traditional law of the sea (*hukum laot*). The management of Aceh's marine and fishery resources to date still involves the Acehnese customary institution, the *Panglima Laot* Institute. *Panglima Laot*, in its implementation, applies the values and concepts of local wisdom. The concept of local wisdom is still being maintained. In managing the coastal and marine environment, *Panglima laot* adheres to the customary law of *laot*. *Laot* customary laws are standard rules preserved and maintained by the fishing community to keep order in fishing and fishermen's lives in coastal areas. In carrying out its functions, *Panglima laot* has duties, including coordinating and supervising every fishing effort in the sea, guarding and supervising *uteun bangka* (mangrove forest), *uteun pasie*, *uteun aron* (pine forest), *neuheun* (pond), and *lancang sira* (salt fields), *bineh pasie* (beachside). *Uteun Bangka* (mangrove forest) is a buffer zone for life on the coast. The values of Indonesian local wisdom that have been passed down by our ancestors must continue to be preserved. Local wisdom possessed by each region can support environmental rehabilitation efforts after a disaster such as a tsunami disaster.

Aceh is a disaster-prone (BPBA 2019) area requiring early knowledge related to disaster mitigation that the community understands. The people of Aceh have the values of local wisdom, which, when examined and interpreted, have taught the community to be ready to face disasters with the traditional house of Aceh as a cultural heritage called *Rumoh Aceh*. Hairumini (2016) analyzed the values of *Rumoh Aceh* for earthquake and tsunami disaster mitigation. The values of *Rumoh Aceh* are found in the components of building form and ceremonies. The knowledge, attitudes, and behaviors of local wisdom in *Rumoh Aceh* have taught the community to adopt these values in building their houses, as part of caring for cultural heritage, disaster mitigation, building families and social systems (Elliott 2014). The construction of a housing structure that follows the Aceh traditional house (*Rumoh Aceh*) can be seen in Fig. 9.3.

Together with various related parties, the Aceh government wants to make post-tsunami Aceh a green province through a rehabilitation and reconstruction program. Efforts to realize "Green Aceh" and the development carried out must be environmentally sound and based on Acehnese local wisdom. Examples of the application of Acehnese local wisdom are such as the application of Acehnese customary law in managing marine natural resources and fisheries. The environment in the concept of Acehnese customary law is a gift from God that has strategic value for human life and other living beings. Wise and prudent environmental management has been practiced for a long time and generations.

Mangrove forest plants have various functions, including breeding grounds for multiple types of fish and shrimp, preventing the infiltration of seawater into the land, and resisting abrasion. In some places, such as in Aceh Besar and Aceh Barat District, there is a tradition to plant a mangrove tree, and who plants the tree can benefit from that plant. However, due to uncontrolled management, mangrove tree planting continues to expand, so ownership is no longer clear. Mangrove trees that are already large are cut down by their owners, converted into *neheun* (ponds) to be used as a place for cultivating shrimp, oysters, and fish. This condition causes mangrove forests belonging to *ulayat* communities (the communities' area) to decrease. As a result, mangrove preservation for coastal ecosystems' protection becomes difficult because of the transfer of ownership of mangrove forests in the local area.

Uteun Aroen (pine forest) is a coastal buffer zone that is dominated by pine trees. Based on the fishing community's experience, information is obtained that the waters near the coast with many pine trees are very popular with fish groups such as snapper, grouper, and mollusk groups (shellfish). Interest in fish and other marine animals is also thought to be related to the temperature factor's condition around the coastal area where pine trees are growing. After the tsunami, cypress trees were again planted by the community because they were believed to protect coastal areas and bring benefits to the surrounding animal life. *Uteun pasie* (coastal forest) is a canopy area of forest trees that grow along the coast. *Uteun pasie* is a protected forest area for the benefit of maintaining environmental balance in coastal areas. Aceh coastal communities have local wisdom in regulating coastal areas, including through the protection of green lines in the form of trees along the coast, better known as green belts. In the past, the green lane was like *bak aron* (pine tree) protected by traditional institutions. Logging like *bak aron* in the *bineh pasie* (seashore) will be subject to customary sanctions without clear reasons. At this time, the customs regarding the care and protection of *bak aron* and other trees are starting to get less attention and care.

Bak aron and *bak mee* (tamarind tree) that survived the tsunami waves were cut down for reconstruction purposes such as making road expansion. According to customary stakeholders' experience, *bak aron* has a relatively good ecological function to create a balance in the environment in coastal areas. Its high adaptability to coastal climates and seawater makes *bak aron* tubs survive even though the land where it grows is inundated by seawater at high tide. From an economic viewpoint, the green line like *bak aron* attracts certain fish species to approach the coastal area.

This condition is, of course, very beneficial for fishers because they can catch fish in the nearby waters. Another benefit of planting *bak aron* is that it can strengthen cliffs or soil ties along the coast where it grows.

The facts in the field indicate that the *bak aron* can survive in the tsunami puddle. Baked *bak aron* and other tree species can still be seen around Lhoknga Beach, and Lampuuk Beach, Aceh Besar District, was the worst hit by the tsunami. Apart from the *bak aron*, according to the local community, there are several other types of plants that are good and important to be planted in the buffer zone (seashore), such as *bak bunot* (*nyamplung/Callophylum inophyllum*), *bak seukee* (pandanus), and *bak Bangka* (mangroves). Society can plan on dry coastal soil or in tidal areas. In efforts to maintain the preservation and abundance of fish in coastal areas, the *Panglima laot* regulates fishing time, catching procedures, and use of fishing gear through local sea customs. Provisions or prohibitions held in *laot* customary law are still useful in fishing communities. The requirements or prohibitions of customary sea law are as follows:

1. Fishermen are not allowed to destroy the habitat of fish and other biotas, through bombing, poisoning, anesthesia, electricity, taking coral reefs, and any other means.
2. Fishermen must not destroy and cut down timber trees on the beach, such as pine, pandanus, ketapang, mangroves, and other trees that live on the coast.
3. Fishermen are prohibited from fishing and destroying protected marine life such as turtles and dolphins. Using nets in coral reef areas or fish spawning areas is not permitted. There are fishing arrangements that have tags.

Every aspect of activities at sea is regulated by traditional sea law. These regulations are flexible according to the developing legal needs of the fishing community. Examples of new rules that have been mutually agreed on are the prohibition of fishing using trawl-type fishing gear (trawl), explosives, electricity, and poison and a ban on disposing of waste such as used oil into the sea. The rule must protect the sea's potential in such a manner to achieve sustainability not only by strict regulation from destructive behavior but also in favor of the local wisdom of *Adat laot* and *Panglima laot*. Local people view environmental sustainability as something that integrates nature with values (Abdullah et al. 2018).

9.7 Conclusion

Every region in Indonesia has different biodiversity and cultural diversity. Biodiversity, which is a natural resource, forms the basis for determining the cultural characteristics of the community in a particular area. Areas with higher biodiversity have greater cultural diversity (Hong 2011). The human ability to respond to global environmental changes, such as the impact of natural disasters, is influenced by the development of human knowledge in utilizing biological resources. Environmental change conditions such as those caused by natural disasters require time in efforts to adapt and rehabilitate the environment based on local wisdom. Therefore, people affected by natural disasters such as the tsunami in Aceh need to build an environment-friendly lifestyle and revive themselves in optimizing the advantages and disadvantages of the affected areas, such as the uniqueness of the coastal ecosystem and its cultural uniqueness, which is a value of its local wisdom. Aceh House architecture, replanting of Aceh forests, cultivation of marine biota and terrestrial animals, and coastal and marine management by the sea commander based on Aceh's sea law are part of post-tsunami rehabilitation in Aceh based on local wisdom. This local wisdom grows and develops in the local scope and adapts to the needs and circumstances of certain communities. Local wisdom must be continuously preserved, developed, and passed on to the next generation through technological innovation and education as an effort to mitigate the impact of natural disasters.

Acknowledgments We want to express special thanks to our friends who gave us the golden opportunity to write the book chapter on Conserving Biocultural Landscapes in Malaysia and Indonesia for Sustainable Development. We are also very thankful to the Biology Department of the Mathematics and Natural Sciences Faculty for supporting us in finishing this book.

References

Abdullah MA, Arifin A, Tripa S (2018) Panglima Laot: his legacy and role in conserving marine resources in Aceh, Indonesia. In: Abdullah NFLB, Esa NB (eds) SHS web of conferences, vol 45. EDP Sciences, Les Ulis, p 06003

Bhandari MP (2018) Climate change science: a historical outline. Adv Agr Environ Sci 1(1):5–12. https://doi.org/10.30881/aaeoa.00002

BMKG (2019) Indonesian tsunami catalog 416-2018. Climatology and Geophysics Council, Jakarta. (In Bahasa)

BPBA (2019) Aceh province disaster management plan 2020-2022. BPBA, Aceh. (In Bahasa)

BRR & Patners (2006) Aceh and Nias, two years after the tsunami, Banda, BRR

Cochard R et al (2008) The 2004 tsunami in Aceh and southern Thailand: a review on coastal ecosystems, wave hazards, and vulnerability. Persp Plant Ecol Evol Syst 10(1):3–40

Díaz S et al (2018) Assessing nature's contributions to people. Science 359(6373):270–272

Elliott CL (2014) At home in the post-tsunami landscape? A case study of post-disaster housing in Aceh, Indonesia. Thesis. University of Tasmania, Hobart

Erianjoni (2016) The formulation of natural disaster-based local wisdom values after tsunami disaster in Aceh Jaya District, Aceh Province, Indonesia. Res Hum Soc Sci 6(16):120–124

Hairumini H, Setyowati DL, Sanjoto TB (2016) Local wisdom of Aceh traditional house as a heritage culture for earthquake and tsunami disaster mitigation. J Edu Soc Sci 5(1):89–96

Hong SK (2011) Bio-cultural diversity and traditional ecological knowledge in island regions of Southwestern Korea. J Ecol Field Biol 34(2):137–147

Martinez-Reyes J (2012) Bio-cultural diversity conservation: a global sourcebook. Ethnobiol Lett 3: 61–62. https://doi.org/10.14237/ebl.3.2012.46

Measey M (2010) Indonesia: a vulnerable country in the face of climate change. Glob Major E-J 1(1):31–45

Meilianda E et al (2019) Assessment of post-tsunami disaster land use/land cover change and potential impact of future sea-level rise to low-lying coastal areas: a case study of Banda Aceh coast of Indonesia. Int J Disaster Risk Red 41:101292

Mutaqin BW, Lavigne F, Hadmoko DS, Ngalawani MN (2019) Volcanic eruption-induced Tsunami in Indonesia: a review. IOP Conf Series: Earth Environ Sci 256:012023. IOP Publishing. https://doi.org/10.1088/1755-1315/256/1/012023

Noor YR, Khazali M, Suryadiputra INN (2006) Guide to introducing mangroves in Indonesia. Ditjen PHKA: Wetlands International, Indonesia Program, Bogor. (In Bahasa)

Poisson B, Oliveros C, Pedreros R (2011) Is there a best source model of the Sumatra 2004 earthquake for simulating the consecutive tsunami?: The best source for Sumatra 2004 tsunami?'. Geophys J Int 185(3):1365–1378

Rotty AN (2018) The role of World Wide Fund for Nature (WWF) in the reconstruction and rehabilitation of aceh post-tsunami. eJournal Ilmu Hubungan Internasional 6(3):909–922. (In Bahasa)

Suryawan F (2007) Study of vegetation and physical conditions of coastal area and conservation efforts in Nanggroe Aceh Darussalam: 121 (In Bahasa). Biodiversitas J Biol Divers 8(4):262–265

UNEP (2007a) Environment and reconstruction in Aceh: two years after the tsunami. UNEP, Nairobi

UNEP (2007b) After the tsunami: coastal ecosystem restoration. UNEP, Nairobi

UNESCO (2017) Basic texts of the 1972 world heritage convention. UNESCO, France

Wibisono ITC, Suryadiputra IN (2006) From mangrove/coastal ecosystem restoration efforts in Aceh since the Tsunami. Wetlands International, Indonesia Program, Bogor, p 96

WIIP & WWF Indonesia (2008) Coastal ecosystem rehabilitation in Jaya district, Aceh Jaya Regency (In Bahasa)

Weaving the Semelai Knowledge-Practice-Belief: Traditional Ecological Knowledge and Farming in Tasek Bera, Pahang State, Peninsular Malaysia

10

Zanisah Man and Sharina Abdul Halim

Abstract

The importance of indigenous knowledge is widely acknowledged in the reclamation, revitalization, and renewal of knowledge systems, culture, lands, and resources. Moreover, indigenous ways of knowing have continued to persevere despite the historical and ongoing processes of colonization and state-sanctioned efforts and programs and continue to be reclaimed and sustained by indigenous peoples across diverse global regions. Using a qualitative ethnographic research approach, this chapter examines the importance of traditional ecological knowledge among the Semelai (in determining the fertility of their land), the ways in which they sustain their knowledge of plants over generations, and how external factors contribute to various changes in their ecosystem. The main findings from a preliminary analysis of Semelai ecological knowledge are two-fold—the need for an approach in understanding knowledge-practice-belief within the Semelai community, and the process of inquiry/systematic analysis within indigenous ecological knowledge in order to comprehend the concept of sustainability. The way forward to conserve Semelai ecological knowledge is to increase the discourse on the same, which plays an important role in the community's ability to recognize and reclaim their rights and to exercise their knowledge of sustainable development for present and future generations.

Z. Man (✉)
Anthropology and Sociology Program, Faculty of Social Sciences and Humanities, Universiti Kebangsaan Malaysia, Bangi, Selangor Darul Ehsan, Malaysia
e-mail: zanisahman@ukm.edu.my

S. A. Halim
Institute for Environment and Development (LESTARI), Universiti Kebangsaan Malaysia, Bangi, Selangor Darul Ehsan, Malaysia
e-mail: sharinahalim@ukm.edu.my

S. A. Abdullah et al. (eds.), *Conserving Biocultural Landscapes in Malaysia and Indonesia for Sustainable Development*,
https://doi.org/10.1007/978-981-16-7243-9_10

Keywords

indigenous epistemologies · Orang Asli · swidden farming · traditional ecological knowledge · Tasek Bera · indigenous peoples

10.1 Introduction

Knowledge is a powerful tool in the crafting of ideas and the construction of reality. Traditional knowledge, however, is often dismissed because its values are considered to be exclusive to a particular society and may not be supported by scientific evidence (Siti et al. 2009). This has resulted in an increased sidelining of traditional knowledge, with traditional practitioners abandoning their knowledge for more "modern" knowledge systems that may not necessarily be of benefit to the community (Masipiquena et al. 2000; Laudine 2009). Still, there exists a lack of understanding and acknowledgement of the vast and long-standing knowledge of indigenous peoples in relation to their interaction with their environments. The ways in which indigenous communities respond to changes are an integrated part of many indigenous and local cultures (Siahaya et al. 2016; Toy 2015), and the combination of knowledge, practices, and cultural traditions is usually passed down from generation to generation. Such knowledge and practices often relate to land use and the functioning of biodiversity and their ecosystems (Parotta et al. 2016). The recognition of the contribution of indigenous peoples and their knowledge as stated in Agenda 21 is a critical factor in the quest for a sustainable future. There are numerous references to indigenous knowledge, or what is commonly known as traditional ecological knowledge (TEK), such as stated in the Preamble of Articles 8 and 10 of the Convention on Biological Diversity and the United Nations Declaration on Indigenous Rights (UNDRIP 2007).

TEK represents a body of knowledge that has been acquired over thousands of years of direct human contact with their environment (Berkes 1993), which includes knowledge acquired by indigenous and local peoples through direct contact with the environment. Within the perspective of a traditional ecological framework, traditional knowledge holds a certain value that is deeply rooted within the community's system-of-knowledge analysis. This includes the argument that knowledge belonging to traditional society may include scientific inquiry, although approaches in evaluating and understanding certain observations may differ (Laudine 2009). The idea of TEK as a form of science stems from the argument made by Berkes (2008), who insisted that the scientific knowledge we possess today is based on the curiosity-driven inquiry made by traditional peoples and ancient societies. His idea is heavily influenced by Levi-Strauss (1962) beliefs, as contained within Strauss' book The Savage Mind. Berkes (1993) emphasizes that "the native knowledge of natural milieu is firmly rooted in the reality of accumulation of concrete, personal experiences, as opposed to book learning."

Control of their knowledge and ability to manage their lands through traditional systems is an essential part of many indigenous peoples' movements in reclaiming their cultural heritage and revitalizing their communities (Alcorn et al. 2003; Kimmerer 2002; Ross and Pickering 2002; Berkes 2012). UNDRIP's Article 3 states, "Indigenous peoples have the right to self-determination. By virtue of that right, they have freedom to determine their political status and to pursue their economic, social and cultural development" (2007). Indigenous peoples' ability to claim self-control and management rights to their lands is situated in the acknowledgement of TEK as a valid knowledge source. Respect for TEK, the knowledge system that informs and is informed by a community's relationship to land, allows communities the opportunity to possess self-control over their resources and culture (Casson 2015).

In this chapter, the situation faced by the indigenous communities of Peninsular Malaysia—in particular the Semelai community of Pos Iskandar, Pahang state, is discussed, with an examination of the challenges they face in upholding their TEK despite rapid changes in their natural environment, and the economic, political, and cultural changes occurring at various scales and levels—with many of their knowledge systems at risk of disappearing. As indigenous peoples in Peninsular Malaysia (the Orang Asli) are not immune to this threat, there exists a need to identify and document such knowledge for scientific research, so as to enhance contemporary understanding of ecologies within the forest system. In light of this, the study's objective is to explore the importance of indigenous knowledge to the community (in determining the fertility of their land), the ways in which they sustain their knowledge of plants over generations, and how external factors contribute to change. Despite dismissal of their knowledge systems by mainstream society, the Semelai continue their attempts to conserve their knowledge. This preliminary study will also apply the TEK approaches as introduced by Berkes (2008) to the Semelai community.

10.2 The Community

The ethnic label *Orang Asli*, meaning "natural people" in Malay, replaced the term "aborigines" that was used by the British colonial administration (Gomes 2004). *Orang Asli* refers to the indigenous peoples of Peninsular Malaysia. The *Orang Asli*, together with the Malays and indigenous peoples of Sabah and Sarawak (Malaysian Borneo), form a category of Malaysians known as the *bumiputera* ("sons of the soil") who make up 65.1% of the population, with the remaining major ethnic groups consisting of communities of Chinese and Indian descent (http://www.statistics.gov.my/English/pressdemo.htm). The *Orang Asli* comprise 0.6 per cent of the population (JAKOA 2011) and are conventionally divided into eighteen ethno-linguistic subgroups (Juli 1998; Nicholas 2000).

Orang Asli land is coveted by powerful interests for its timber and minerals, for conversion into oil palm or rubber plantations, golf courses, hydroelectric power installations, airports (such as Kuala Lumpur International Airport), and development projects that benefit the majority population. Although the Aboriginal Peoples Act (1974) permits the Orang Asli to collect minor forest products, under the Forest Policy (which began in the 1930s), the Forestry Department reserves regulatory rights. The Act requires traders to obtain licenses to purchase or trade forest products and to pay levies and taxes on commodities. By such means the department is capable of regulating trading and controlling Orang Asli access to forests.

As a consequence of the rapid development in their natural environment, followed by economic and political changes and pressures, the conversion of Orang Asli land in the name of development has also affected their knowledge systems, not only in terms of their ability to practice their knowledge, but also in its passing down to future generations. In order to detail the reasons behind the ongoing plight of the Orang Asli and in explaining their response to their plight, it is vital to not only explore Orang Asli cultures, but also the attitudes and aspirations that shape the government's policies for the Orang Asli (Dentan et al. 1997). Moreover, this paper concurs with the statement made by Wan (1996) regarding the fact that the "Orang Asli are not anti-development." Indeed, the development of policies and fairer laws is becoming increasingly possible, particularly as a result of Orang Asli leaders who are now becoming more knowledgeable in the advocating of land titles, as well as in their ability to express their expectations of the government for their community.

The general thread of these expectations is that the Orang Asli desire development and integration into mainstream society, but with their ethnic identities remaining intact. It is not the intention of this paper to undermine the struggles that the Orang Asli face in their journey to materialize these aspirations, given that there is still a rather long road ahead for them before they become self-sufficient and are able to liberate themselves from poverty (largely due to government policies and laws that are largely not in sync with the needs of the Orang Asli (Idrus 2011)). Such interventions have affected and impacted the way Orang Asli "indigenous knowledge" is passed down from one generation to the next, which includes TEK—which encompasses not only knowledge of the forest, ecology, and wildlife, but also relationships with socio-economic, cultural, and spiritual aspects, as well as Orang Asli identity, which is deeply rooted to their original lands.

10.3 Research Area

Home to an estimated population of 2029 (JAKOA 2011), Pos Iskandar is one of the largest Semelai villages in Tasek Bera, Pahang, Peninsular Malaysia. The name Pos Iskandar in fact originates from a British military base known as Fort Iskandar, established in 1948 among the old Semelai village and farms during the Malayan Emergency of 1948–1960. There are 22 small villages under Pos Iskandar's administration, which are divided into five primary clusters or areas by the Department of Orang Asli Development (also known as JAKOA) for the administration's convenience—Gau-Pelawan, Putat, Bapak, Jelawat, and Kuin/Lengot. These five areas are separately governed by five different *batins* (headmen) although in actual social settings or matters of representation, the five *batins* generally work as a team.

Tasek Bera is the largest natural lake in Peninsular Malaysia. Famous for its vast lake and natural environment, Tasek Bera received international recognition for natural conservation in the 1990s and was nominated as a Ramsar site in 1994 by the RAMSAR Convention on Wetlands. Originally known as The Convention on Wetlands or "Ramsar Convention," RAMSAR is an intergovernmental treaty that embodies the commitments of its member countries to maintain the ecological character of wetlands of international importance, and to plan for the "wise" or sustainable use of all wetland areas in their territories. According to Gharibreza et al. (2013) the lake has a total catchment area of 593.1 km^2, with an estimated 340 km^2 of the original tropical rain forest cover having been cleared since 1972 to accommodate oil palm and rubber plantations. An assessment by Lam et al. (2018) using multi-temporal remote sensing data and GIS analyses from 1990 to 2017 has shown rapid expansion of land use and land cover surrounding the lake area—particularly in the inward expansion of palm oil plantations that have eliminated the previously existing natural forest and rubber plantation bordering the Semelai settlement. Unfortunately, changes surrounding the lake environment have resulted in the disruption of essential ecosystem goods and services that benefit the low-income Semelai community—many of whom are reliant on subsistence economy.

10.4 Methods

The research is based on ethnographic data collected from approximately two years of study among the Semelai community of Tasek Bera, Pahang State. The first phase was collected between December 2012 and July 2013; due to time constraints, the second phase was conducted between September 2014 and January 2015 on an "on and off" basis. To successfully document local ecological knowledge, long-term engagement (Butler 2004) and learning from (or with) the community is needed. Therefore, the two most prominent techniques in ethnographic research (in-depth interview and participation observation) were employed in this study to understand and explain TEK approaches among the Semelai. In-depth interviews were carried out with selected leaders (headmen) and key informants (the shaman, the medicine man, and elders) who are recognized by the community as being knowledgeable and highly respected. Interviews were mainly unstructured and open-ended. Moreover, observation and participation with members of the community as they go about their daily lives is crucial to understanding their culture and its significance to the community.

In order to observe changes in TEK practices among the Semelai of Tasek Bera, 100 survey questionnaires were distributed to respondents of the two villages of Putat and Jelawat, which are located within Pos Iskandar. These two villages represent 26% of the estimated 2029 Semelai population of Pos Iskandar. Putat is considered the most densely populated village with an estimated 410 residents, whereas Jelawat comes in third after Kuin/Lengot with an estimated figure of 124 inhabitants. Economically, the villagers of Putat rely on rubber-tapping as their main source of income, supplemented with seasonal activities such as fishing and the collection of forest produce. Although rubber-tapping remains an important economic activity that is carried out by most villagers in Jelawat, certain families are still engaged in forest-related activities such as the collection of *agarwood*, as well as herbs and medicinal plants, and the production of handicrafts for cash. In comparison to the villagers of Putat, the Jelawat area (closely located to secondary forests) has indirectly contributed to the diverse economic activities carried out by the area's people—which includes active engagement in swidden activities. The villages' distance to nearby forests and involvement in swidden farming are important considerations toward understanding TEK practices and changes in the same among the indigenous Semelai community.

10.5 Relationship between Traditional Ecological Knowledge and the Semelai Community

The idea that there is an epistemological or substantive distinction between indigenous knowledge and other types of knowledge (western, scientific and non-indigenous) has been, quite rightly, debunked (Agrawal 1995). As Gray (1995) argues—the term indigenous is more imperative than descriptive, referring to a quality that emerges in the course of struggles over rights to territories, resources, and cultural respect. It is within this context that this article examines the relationship between TEK and the Semelai community using Berkes' (2008) framework of TEK (Fig. 10.1). According to Berkes (2008), there are four interrelated layers of traditional ecological knowledge. The first includes local taxonomic knowledge of flora, fauna, soil, and landscape. The second level identifies the functional relationship between resource management systems and local environmental knowledge. The third includes social institutions, codes, and rules that are

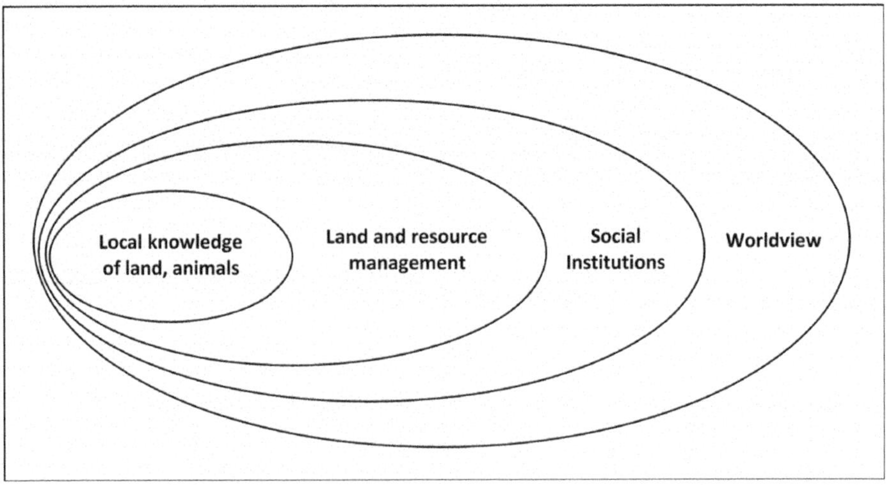

Fig. 10.1 Traditional Ecological Knowledge (Berkes 2008)

in-use and finally, the fourth level is the "worldview" that shapes environmental perception.

The usefulness of traditional ecological knowledge has been recognized by international and national agencies, NGOs, donors, officials from various governmental departments, as well as academics—but not quite in a manner that emphasizes the issue of rights. The potential contribution of TEK to resilience-building in social-ecological systems has gained growing attention in the context of accelerated global change and generalized ecosystem service decline (Turnhout et al. 2012). Throughout history, communities maintaining tight links to ecosystem dynamics have developed knowledge, practices, and institutions to accommodate recurrent disturbances toward securing their livelihoods (Berkes et al. 2003; Balilla et al. 2012). As it co-evolves with ecological and social systems, TEK can strengthen the capacity of human societies to deal with disturbances and maintain ecosystem services under conditions of uncertainty and change (Colding et al. 2003; Berkes and Turner 2006). However, over the last two decades, major developments are reshaping societal perceptions regarding the fate of TEK (Gómez-Baggethun et al. 2013). Firstly, in recent years, researchers have been updating their perceptions of TEK's ability to adapt to change. Until recently, TEK was largely perceived as a vestige of the past that held—at best—folkloric interest, and which was bound to disappear with increasing economic development.

Secondly, new perspectives on TEK's adaptive nature have favored increasing recognition of the value of such knowledge within environmental policies (Reid et al. 2006, Turnhout et al. 2012). International policy processes such as the United Nations' Declaration on the Rights of Indigenous Peoples 2007 (Article 31) and the Convention on Biological Diversity 1992 (Article 8) have encouraged national governments to recognize and protect TEK for the conservation and sustainable use of biological diversity, as well as to promote its wider application in resource management and biodiversity conservation.

As for the Semelai community, and similarly elsewhere such as in the Lindu community of Indonesia (Li 2000), TEK memory-carriers are threatened by the processes of land conversion (Rafaai et al. 2020), resettlement and agricultural industrialization (Abdul-Karim and Hashim 2010), and technology (Friendship and Furgal 2012). Loss of TEK-related practices continues to result in the associated erosion of biodiversity and regulatory ecosystem services (Hariyadi and Harmoko 2015). The introduction of palm oil at the fringe of Semelai territories since the 1970s has resulted in considerable knowledge erosion due to the loss of flora and fauna that are important to the Semelai's construction of traditional housing structures, as well as for medicinal and ritualistic purposes. Certain forms of knowledge associated with hunting-gathering, such as awareness of various yam varieties, poisonous plants, and wild edibles fruits, or even know-hows of different traps, and identification of animal dung and footprints, are slowly vanishing. Detached from the forest-learning environment of their communities, the younger generation may not be able to fully understand the rationale behind ancestral TEK knowledge due to their quest for secular education outside Semelai settlements. The process of inquiry on the epistemology of Semelai ecological knowledge is largely still in its infancy with regard to its documentation. As such, this article provides a preliminary analysis that seeks to understand existing Semelai ecological knowledge by relating it to Berkes' (Berkes 1999, 2008) knowledge-practice-belief framework—first by explaining its epistemological understanding (the *smicdec* and *gohop* concepts that reference the Semelai worldview), followed by an interchangeable discussion of land and perspectives in resource management (via the *rasik* ritual), as well as the social institution (the community) that uphold said forms of knowledge.

10.6 Semelai Ecological Knowledge

Semelai identity and livelihood have traditionally centered on swidden farming for their subsistence needs—in addition to fishing and hunting within their forest territories. The community's sentiment with regard to their subsistence practices is best-described by Gianno and Bayr (2009) "....the Semelai proudly identified themselves as rice cultivators who made large swidden, preferably in primary forest," indicating the community's dependency and vast knowledge of paddy and traditional farming. A further analysis by Gianno and Bayr (2009) from Williams Hunt's aerial photograph collections (spanning from the 1940s to the 1960s) unfolds the scattered localities of Semelai settlements that surround Tasek Bera to the downstream areas of Sungai Serting, Sungai Teriang, and the Sungai Bera watersheds. These aerial photographs displayed substantial numbers of paddy and tapioca trees across their scattered farms—patterns that point to the Semelai's excellent agricultural rotation system, which works in tandem with the natural support provided by vast forest cover in the sourcing of housing supplies, water retention, and the considerably large space available for essential trees to flourish in proximity with Semelai settlements, which facilitates the presence of wild animals within Semelai territories.

In the past, the practice of paddy planting was sufficient to sustain the community's subsistence needs. For example, the abundance of food provided by swidden farming creates occasions for families to celebrate *Nenggulung* day—which marks the end of the paddy season and the sharing of the year's harvest with family members. This feast, although not conducted in large crowds such as found among other indigenous harvest festivals such as *Gawai* or *Kaamatan* in Malaysian Borneo, signified a family's ability (knowledge) to maintain farm and good crops. Given the considerable distance between settlements and transportation infrastructure, the event creates a platform where family members can connect with each other, and, perhaps most importantly, exchange knowledge of the area's suitability for farming (such as soil conditions)—a trans-generational informal learning process for both adult and younger Semelai community members.

In Semelai culture, the knowledge and ability to determine the suitability of a given area for agriculture or settlement reasons is fundamental, as it may prevent potential harm to families from both physical and spiritual beings. This fundamental knowledge actually lies within the individual's ability to grasp the understanding behind the concepts of *smicdec* and *gɔhɔp* in determining the level of coolness of a given area, and in the search for ideal living spaces. This fundamental concept will be used to discuss the Semelai ecological knowledge as an important component of their belief system or worldview.

10.7 The Concepts of Coolness (*Smicdec*) and Warmness (*gɔhɔp)*

Smicdəc is an important survival concept that stems from the Semelai's cosmological view. Similar to the concept described by Hood Salleh (Hood 1978) in the *bəbəlian* ritual—coolness (*sədəc*), it references the condition when malevolent spirits are not harassing a person's soul—thus a state of desirable balance as perceived by Semelai culture. The land (*ʔatɛ*) is *smicdəc* when its dwellers are free from malevolent spirits capable of harming them both physically and spiritually. Once the state of *smicdəc* is obtained, it creates a desirable social equilibrium that may be measured by health and fortune (abundant food and wealth) it brings to settlers.

The opposite characteristic of *smicdec* is "warmness" (*gɔhɔp*). The community recognizes the land as being *gɔhɔp* through stories mentioned by shamans who warns them to avoid certain sacred areas. Simultaneously, the elders played an important role in transmitting knowledge of sacred areas to the younger generation via oral history—thus shielding the community from unnecessary danger. However, there are certain places that were not mentioned by shamans in their oral history as sacred, but which do not necessarily mean *smicdec*. To determine a land's *smicdec* levels, a ritual called *rasik* must be conducted. Within the traditional ecological knowledge continuum as stated by Berkes (2008), it can be said that in the Semelai worldview of ecological knowledge, the *rasik* ritual is embedded within the Semelai agricultural calendar—mentioned by Hood Salleh (Hood 1978) as *ring bri* (locating a forest).

10.8 Ritual to Determine Suitability of Forest Land

Rasik is a ritual conducted to determine whether an area is suitable for clearing prior to setting up a farm (*doʔoh*). The ritual will determine the land's level of coolness (*smicdec*), which is crucial for either crop planting (such as paddy) or the cultivation of edible fruit trees. The ritual can be carried out without the presence of a shaman or *bomoh*. Anyone from the community may perform *rasik* as long as they are willing to learn its methods from elders or a community member with a deeper understanding of the ritual. To perform *rasik* for a land (usually forest land), one needs to understand three important steps within the confines of taboo (Fig. 10.2). Firstly, the taboo right before they step out of their house and into the forest; secondly, the taboo while the individual is in the forest; and lastly, the taboo once they return home from the forest.

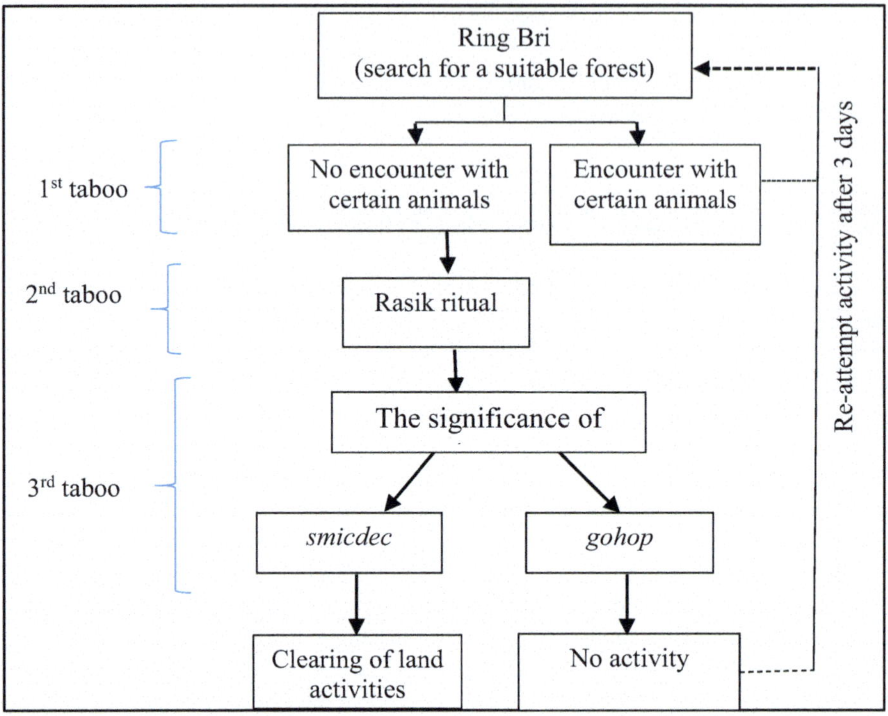

Fig. 10.2 The *rasik* ritual

The first taboo (prior to leaving the house) is that the individual does not inform others of his intention with the exception of a few friends (three to four persons) who will accompany the individual into the forest. In the event that the person encounters any type of snake, or a particular bird known as "*cɛp*" after they step out of their house for their journey, they would have to immediately cancel their plans and re-attempt it after three days. Should the individual not encounter any such animals along the way, they may proceed with their intentions as planned.

Once he arrives in the area, he will begin the ritual (the second taboo) (see Fig. 10.3). There are at least five different types of *rasik* that can be conducted (depending on the individual) but only two types (those normally practiced) will be highlighted here with regard to TEK practices within Semelai ecological knowledge. Before the ritual actually begins, the individual will need to search for a specific tree that the Semelai call *tənbon* (*Baccaurea parvifolia*) within the chosen forest area. Once the *tənbon* tree is found, he clears a small area of about five to ten square feet. The act of clearing the area surrounding the *tənbon* tree is known as *chɒr ʔatɛ*. Once the area has been cleared, the person searches for small trees to be made into sticks that are cut to the length of his arm span (*mədəpak*) (Fig. 10.3). This is followed by the making of a small hole beside the *tenbon* tree with the same stick used earlier for measuring. Once the hole has been dug, he performs the following act (Fig. 10.3);

Fig. 10.3 Rasik ritual performed by a Semelai indigenous man from Tasik Bera (Photo by Z Man)

bringing the stick toward his chest (in the center), holding the stick horizontally with both hands, blowing into the space between his hands (with his mouth) and stroking (*saʔap*) the stick gently using his left hand toward the far left end, before placing his hand back to the center; this is followed by his right hand, which gently strokes the stick toward the right, with the same action repeated to the count of seven. Once he finishes, he will utter the words "*tunyuk rasik*" (show me [the] *rasik*) and at the same time insert one end of the stick into the hole in the ground, after which he measures the stick once again using his arm span. Should the measurement be a few centimeters shorter than the previous arms span, this would be an indication that the land is not meant to be cleared. Shorter measurement implies a shortened life span for the dwellers (death) should one insist settling in the area. Otherwise, if the stick is longer by a few centimeters than the measurement done previously, the land is considered suitable for farming. This is followed by placing the stick inside the hole next to the *tənbon* tree, and leaving it there (leaning against the *tenbon* tree).

The second type of *rasik* is conducted by making a hole (the size of a wrist) in the ground, clearing the area next to the hole and leaving the hole for three days. Should the hole remain empty after three days, this indicates a good sign, but if the hole has leaves in it (with young or fallen yellowish/brownish leaves), the land is not suitable for farming or living. Young leaves mean that a youth/young child in the family will die should the person conducting the ritual decide to proceed with his plans of clearing the land and inhabiting the area, while yellow/brownish leaves indicate that the land will claim the life of an adult. A rather similar agricultural ritual is conducted by the Katu in Vietnam to determine the suitability of their "field" (*hare*) and "forest" (*krung*) before the clearing of a particular area for agricultural activities. The Katu will perform extensive ritual procedures before moving to a certain area

(Arhem 2009). Such rituals are important in informing the community regarding the disposition of spirits within the area, and the ritual is a form of "permission-seeking" prior to clearing the forest and working the land. The Katu also consider a number of factors such as the condition of the soil (good or bad), the status of the land (previously cleared or otherwise), and the spirit residing in the area (benevolent or malicious) before they decide to clear forests in a particular area. These procedures are vital in determining the suitability of the place.

In the Semelai community, dreams are important in interpreting the will of the spirits—the third taboo. Therefore, once the ritual has been completed and the person returns home to sleep at night, the dream one encounters on the same night would indicate whether the area is suitable to farm in the future. Dreams consist of either bad or good dreams—if in one's dream, the individual kills animals, sees paddy, and a strong water current, this would translate as a bad dream. The killing of animals signifies death, while paddy foretells the shadow of death to the person himself or to his family; strong water current, however, indicates blood and injuries. Subsequently, the individual should not proceed with his intention of creating a farm in the area. If he dreams of playing with water, or sees a funeral or grave, and a small child—this would indicate a good dream. Playing with water signifies the area's "coolness" (*smicdec*), while a funeral or a grave means that the land is fertile and will produce bountiful paddy/rice for the family; a small child appearing in the dream would mean there will be an abundance of paddy/harvest for the family for years to come.

After one experiences a good dream, he should wait for three days before returning to the area and initiating land-clearing activities. Once the area is cleared and the burning process (*chɔr doʔoh)* completed, he would then need to perform another ritual known as *rasik baba* should he intend to plant paddy in the area. *Rasik baba* (paddy ritual) differs from *rasik ʔatɛ* (the land ritual) in terms of methods and the use of "*nəngulang*" (*tenbon* plant), tumeric (*nəris)*, black sugarcane (*bos hitam*), and paddy seeds he intends to plant while performing the ritual.

The above ritual of determining land suitability portrays the Semelai's ecological understanding and attachment to their surrounding environment. In order to understand Semelai knowledge-practice-belief approaches, an analysis of each layer of Berkes' TEK approach should be understood. Based on the TEK continuum, the first layer of analysis consists of local and empirical knowledge of animals, plants, soils, and landscape—the *cɛp*, the *tenbon* tree, and the land. The ritual of observing soil and landscape suitability for paddy planting is carried out through local knowledge and a set of conducts—which includes the observation of "arm measurement" and the "hole" [second layer of analysis]. The ritual's implied sanctity by the shaman and elders is representative of the knowledge institution within the Semelai community in ensuring that the ritual's processes are conducted accordingly—the dream, its interpretation, and dangers [third layer of analysis]. Finally, by following these layers of processes, the epistemological understanding of the *smidec* and *gohop* concepts is strengthened within the Semelai belief system (world view). However, the *rasik* ritual is presently being contested due to changing agricultural patterns in response to the cash-crop economy. Similar issues are being encountered by other indigenous peoples such as the Iban and Bidayuh in Sarawak of Malaysian Borneo (Cramb 2015), the Bulang in Xishuangbanna, China (Yin et al. 2015), and the Batak of Palawan Island, Philippines (Novellino 2015)—many of whom have lost their TEK knowledge—the diversity of land use and cropping, as well as the authority of headmen and elders in the selection of land and the conducting of rituals, much of which is due to the shift from swidden agriculture to mono-crop plantations (Yin et al. 2015).

10.9 Contested Knowledge

The government's effort to eradicate slash-and-burn cultivation (i.e. swidden farming) has resulted in an increased number of Orang Asli communities shifting from subsistence economy to cash-crop production. The change from subsistence economy (swidden farming) to cash-crop production has significantly transformed the economy of the Semelai and those of Orang Asli communities in Peninsular Malaysia as a whole (Gianno and Bayr 2009; JAKOA 2011). It has also changed the views of the Semelai toward their lands from a cultural perspective and reduced the importance of rituals and taboos in the carrying out of land-clearing activities. In Putat (Table 10.1), 20 respondents (40%) mentioned that they conducted the *rasik* ritual prior to the clearing of new lands, of which most were elders born in the pre-1970s. Another 21 respondents have never performed such rituals either because they had already been carried out by the community's elders (father or father-in-law), or because the cleared land was an abandoned old farm inherited from their ancestors (*do-doh padey*). While in Jelawat, the percentage of respondents who had performed these rituals was higher (62%) compared to Putat, as villagers in the former were still involved in paddy-planting activities. Abandoned farms within this context are based on the community's years-long observation of the land, which takes into account the age of trees and their growth. However, the ritual must be conducted if such abandoned old farms are approximately 15 years and above. As previously mentioned, this knowledge is similar to those of the Katu of Vietnam, in which farmers first perform a series of divinatory procedures to seek approval from the spirits prior to inhabiting the land. The rituals do not have to be repeated by the family at a later point, as the spirits would already be acquainted with the family and its descendants (Arhem 2009).

Paddy-planting activities have reduced significantly among the Semelai community since the 1970s (Gianno 1990; Hoe 2001; Gharibreza et al. 2013; Gianno and Bayr 2009) due to the introduction of rubber trees and oil palm plantations. During the span of this fieldwork, only three families in Jelawat had planted paddy for the year (Table 10.2). No one is involved in paddy-planting activities in Putat because of

Table 10.1 Performing of *rasik* rituals during land-clearing activities

	Yes	No	Unsure	Total
Putat	20 (40%)	21 (42%)	9 (18%)	50 (100%)
Jelawat	31 (62%)	10 (20%)	9 (18%)	50 (100%)

Table 10.2 The proportion of hill paddy farming (*baba doʔoh*) in both villages

Village	Yes	No	Total
Putat	0 (0%)	50 (100%)	50 (100%)
Jelawat	6 (12%)	44 (88%)	50 (100%)
Total	6 (12%)	94 (88%)	100 (100%)

Table 10.3 The likelihood of paddy-planting activities in the future

Village	Yes	None	Unsure	Total
Putat	9 (18%)	38 (76%)	3 (6%)	50 (100%)
Jelawat	26 (52%)	21 (42%)	3 (6%)	50 (100%)
Total	35 (35%)	59 (59%)	6 (6%)	100 (100%)

the hard work involved and the diminished availability of land due to rubber and oil palm agriculture. The reason as to why the community in Jelawat still practices paddy planting today is because they do not wish to lose "good paddy seedling" (which is a local tradition) that has been passed down to them for generations, as well as a desire for better-tasting rice (in comparison to store-bought rice). However, even members of this community admit that growing paddy has become a challenging task as fewer individuals are interested in the activity, due to the risk of wild animals and pest attacks. Additionally, additional work is required to maintain a good yearly harvest year.

When enquiries were made to respondents regarding the continuation of paddy-planting activities in the future (Pos Iskandar area), 38 individuals from Putat were of the opinion that the activity will not survive (Table 10.3). Nine respondents from Putat thought that villagers from Jelawat would likely still practice paddy planting due to the vast forest area surrounding the village. In Jelawat, about half of respondents (21) think that the practice of paddy planting will not survive into the future. Several families who used to grow paddy annually have considered quitting the activity and shifting to rubber and oil palm cultivation (which in turn provides the household with better income). In this regard, although the community values their paddy-planting knowledge, changes in the surrounding environment have forced the community to adapt to new agriculture and economic demands. The same concern was raised by several communities who attempted to maintain their identity through the planting of organic rice using traditional production methods—without losing their plots of land to commercial agriculture such as hybrid rice and monocrops (Yamyin et al. 2020; Yin et al. 2015).

Paddy planting is a culturally important activity that makes a significant contribution to the Semelai's sense of attachment and connection to their land (the concept of *smicdec*). In this context the community is constantly aware of the spirits surrounding them and learns to cohabit with them in the same place and space. This creates a set of values that are sensitive to the environment, due to the belief that change/s are signs, or a way of communicating with the spirits from the unseen world or environment. Arhem (2009) refers to this relationship as a mutual obligation between spirits and the human to warn each other of their existence. The TEK continuum acknowledges this human-spirit connection in its explanation. However, when paddy-planting culture reduced in importance, the cultural aspects and ritual practices behind the concept of *coolness of the land* became less significant, as people began looking for other alternatives (the purchase of food and cash crops) to survive; subsequently, the values and knowledge behind this ritual will gradually be forgotten.

The above analysis indicates that when the environment is disturbed or altered, the existing knowledge-practice-belief system is no longer compatible with the current scenario. This is due to the inability of TEK practitioners—as compared to western science—to use controlled experiments, collect synchronic (simultaneously observed data), and to use quantitative measures in its inquiries (Berkes 2008). Although certain researchers have successfully explored the benefit of traditional ecological knowledge under the discourse of climate change (Colfer et al. 2015; McMillen et al. 2017), including medicinal and health practices (Sinjela and Ramcharan 2005), as well as the risks of increasing environmental exploitation (Friendship and Furgal 2012; Nhem et al. 2018), ignorance toward such knowledge still persists due to contradicting beliefs and interests. Consequently, a different set of inquiries and explanations (using Berkes 2008) is required to educate the mainstream community on the importance of the relationship between humans and the fragile ecological environment (this preliminary research analysis was conceptualized by the Semelai community via their spirits-human interaction) before it is lost due to human detachment from the forest-learning environment.

10.10 Conclusion

The entire *rasik* ritual supports Berkes' (2008) traditional ecological knowledge (TEK) in his primary and secondary layer analysis, which emphasizes resource management systems that requires local environmental knowledge including animals, plants, soils and landscapes, toward the effective application of knowledge. In this regard, the traditional ecological knowledge of the Semelai community is constantly challenged (especially with regard to important ancestral paddy-planting techniques), its processes, and the significance of plants and animals related to the soil's "fertility and abundance" (land). Nevertheless, such inquiries and the understanding of certain values embedded within the *rasik* rituals remain relevant to the community's survival, which makes such knowledge either open to debate or to be accepted as scientific inquiry. This type of knowledge does not subscribe to the prevailing capitalist system, but looks into how the economy can sustain the entire community in a sustainable manner. This chapter offers an explanation for sustainable development using traditional ecological knowledge approaches that could be taken into serious consideration in the discourse on sustainable development—through the adoption of a bottom-up perspective—as the planet becomes increasingly unable to withstand prolonged exploitation. Although this article offers a simple method of TEK analysis within the context of Semelai epistemological understanding, in light of the diversity of indigenous knowledge, and the constant ongoing pressure of land use and forest environments, further systemization of the Semelai's process of inquiry with regard to ecological knowledge is needed.

Acknowledgments This work was supported by the Ministry of Education (MOE) Malaysia under the SLAI Program [grant number KPT(BS)UKM_K017481]. The primary author would like to extend her gratitude to Associate Professor. Dr. Shanthi Thambiah and Professor Dr. Juli Edo from University of Malaya for their insights. The Semelai elders—Aki Men, Andak Tin Tun, Uda Zaleha, Busu Malai, Don and Ayit; in memory of the late three elders who have since passed—Pak Ngah Sipek (also known as Jehudi Badek), Busu Mat Nor, and Paklang Samat. Data for this paper was partly drawn from the primary author's doctoral research.

References

Abdul-Karim H, Hashim AH (2010) The effect of a resettlement scheme on the social-cultural changes of the Temuan community. Procedia Soc Behav Sci 42:362–373

Aboriginal Peoples Act 1954 (1974)

Agrawal A (1995) Dismantling the divide between indigenous and scientific knowledge. Dev Chang 26:413–439

Alcorn JB, Bamba J, Maisun S, Natalia I, Royo AG (2003) Keeping ecological resilience afloat in cross-scale turbulence: an indigenous social movement navigates change in Indonesia. In: Navigating social-ecological systems. Cambridge University Press, Cambridge

Arhem N (2009) In the sacred forest: landscape, livelihood and spirit beliefs among the Katu of Vietnam. SANS Papers in Social Anthropology, Goteborg University, Sweden

Balilla VS, Anwar-McHenry MH, J. M. P., Parkinson, R. M. & Banal, D. T. (2012) Aeta Magbukún of Mariveles: traditional indigenous Forest resource use practices and the sustainable economic development challenge in remote Philippine regions. J Sustain For 31(7):687–709

Berkes F (1993) Traditional ecological knowledge in perspective. In: Inglis JT (ed) Traditional ecological knowledge: concepts and cases. International Program on Traditional Ecological Knowledge and International Development Research Centre, Canada, p pg.1-10

Berkes F (1999) Sacred ecology. Traditional ecological knowledge and resource management, Taylor and Francis, Philadelphia and London, UK

Berkes F (2008) Sacred ecology, 2nd edn. Routledge, New York

Berkes F (2012) Sacred ecology, 3rd edn. Routledge, New York

Berkes F, Turner NJ (2006) Knowledge, learning and the evolution of conservation practice for social–ecological system resilience. Hum Ecol 34:479–494

Berkes, F., J. Colding, and C. Folke. (2003). Navigating social-ecological systems. Building Resilience for Complexity and Change. Cambridge University Press, Cambridge, United Kingdom. http://dx.doi.org/https://doi.org/10.1017/CBO9780511541957

Butler C (2004) Researching traditional ecological knowledge for multiple uses. Can J Nativ Educ 28(1 and 2):33–48

Casson S (2015) Socially-just and scientifically-sound: re-examining co-management of protected areas. IK: Otherways of Knowing 1(2):32–64

Colding, J., Elmqvist, T. & Olsson P. (2003). Living with disturbance: building resilience in social–ecological systems. Pages 163–173 *in* F. Berkes, J. Colding, and C. Folke, editors, *Navigating Social–Ecological Systems: Building Resilience for Complexity and Change*. Cambridge University Press, Cambridge, UK. http://dx.doi.org/https://doi.org/10.1017/CBO9780511541957.011

Colfer, C. J. P., Alcorn, J. B. & Russell, D. (2015). Swiddens and fallows: Reflection on the global and local values of 'slash and burn.' In M. F. Cairns (Ed.), Shifting Cuntivation and environmental change: indigenous people, agriculture and Forest conservation (pp. 62–86). London; New York: Earthscan, Routledge

Cramb R (2015) Busy people, idle land: the changing roles of swidden fallows in Sarawak. In: Cairns MF (ed) Shifting Cuntivation and environmental change: indigenous people, agriculture and Forest conservation. Earthscan, Routledge, London; New York, pp 770–793

Dentan R, Endicott K, Gomes A, Hooker B (1997) Malaysia and the original people: a case study of the impact of development on indigenous peoples. Allyn and Bacon, Needham Heights, MA

Friendship KA, Furgal CM (2012) The role of indigenous knowledge in environmental health risk management in Yukon, Canada. Int J Circumpolar Health 71:1

Gharibreza M, Raj JK, Yusoff I, Othman Z, Tahir WZWM, Ashraf MA (2013) Land use changes and soil redistribution estimation using 137Cs in the tropical Bera Lake catchment, Malaysia. Soil Tillage Res 131:1–10

Gianno R (1990) Semelai culture & resin technology. Connecticut Academy of Arts & Sciences, New Haven

Gianno R, Bayr KJ (2009) Semelai agricultural patterns: toward an understanding of variation among indigenous cultures in southern peninsular Malaysia. J Southeast Asia 40(1):153–185

Gomes A (2004) The orang Asli of Malaysia. IIAS Newsletter 35(10)

Gómez-Baggethun E, Corbera E, Reyes-García V (2013) Traditional ecological knowledge and global environmental change: research findings and policy implications. Ecol Soc 18(4):72. https://doi.org/10.5751/ES-06288-180472

Gray A (1995) The indigenous movement in Asia. In: Barnes RH, Gray A, Kingsbury B (eds) Indigenous peoples of Asia. Association for Asian Studies, Michigan

Hariyadi B, Harmoko D (2015) Benuaron: the fruit gardens of the orang Rimba. In: Cairns MF (ed) Shifting Cuntivation and environmental change: indigenous people, agriculture and Forest conservation. Earthscan, Routledge, London; New York, pp 577–596

Hoe BS (2001) Semelai communities at Tasek Bera: a study of the structure of an orang Asli society. Center for Orang Asli Concern, Subang Jaya

Hood S (1978) Semelai rituals of curing. Unpublished Ph.D dissertation. Oxford University, Oxford

Idrus R (2011) The discourse of protection and the orang Asli in Malaysia, vol 29 (Supp. 1):53, Kajian Malaysia

JAKOA (2011) Pelan Strategik Pembangunan Orang Asli 2011–2015. Jabatan Kemajuan Orang Asli, Kuala Lumpur

Juli E (1998) Claiming our ancestors' land: an ethnohistorical study of the Seng-oi land rights in Perak. Unpublished doctoral dissertation. Australian National University

Kimmerer RW (2002) Weaving traditional ecological knowledge into biological education: a call for action. Bioscience 52:432–438

Lam KC, Man Z, Mahmud M, Jibrillah AM (2018) The remote sensing of land cover dynamics in the Tasek Bera wetland RAMSAR site of Malaysia: implications for sustainability. Proceeding the eleventh international Malaysian studies conference, Bangi, 13–14 august

Laudine C (2009) Aboriginal environmental knowledge: national reverence. Ashgate Publishing Limited, England

Levi-Strauss C (1962) La Pensee Sauvage. Librarie Plon, Paris

Li TM (2000) Locating indigenous environmental knowledge in Indonesia. In: Ellen R, Parkes P, Bicker A (eds) Indigenous environmental knowledge transformation. Harwood Academic Publishers, Netherlands

Masipiquena AB, Persoon GA, Snelder DJ (2000) The use of fire in northeastern Luzon (Philippines): conflicting views of local people, scientists, and government officials. In: Ellen R, Parkes P, Bicker A (eds) Indigenous environmental knowledge and its transformations. Harwood Academic Publishers, Amsterdam

McMillen H, Ticktin T, Springer HK (2017) The future is behind us: traditional ecological knowledge and resilience over tine on Hawai'I island. Reg Environ Chang 17:579–592

Nhem S, Lee YJ, Phin S (2018) The impact of forest resource decline: analyzing forest-related income supplements to reduce income inequality and poverty of the Kouy indigenous people living in Kampong Thom province, Cambodia. J Sustain Forestry 37(2):97–199

Nicholas C (2000) The Orang Asli and the Contest for Resources: Indigenous Politics, Development and Identity in Peninsular Malaysia. Copenhagen: International Work Group for Indigenous Affairs; Subang Jaya, Malaysia: Centre for Orang Asli Concerns

Novellino D (2015) Rice-related knowledge, farming strategies and the transformation of swiddens: among the Batak of Palawan Island, the Philippines. New York: Routledge. In Cairns MF (ed.), Shifting cuntivation and environmental change: indigenous people, agriculture and forest conservation. London: Earthscan/Routledge, pp. 891-918

Parotta J, Youn YC, Camacho LD (2016) Traditional knowledge for sustainable forest management and provision of ecosystem services. Int J Biodiversity Science, Ecosystem Serv Manag 12(1–2)

Rafaai NH, Abdullah SA, Reza MIH (2020) Identifying factors and predicting the future land-use change of protected area in the agricultural landscape of Malaysian peninsula for conservation planning. Remote Sens Appl Soc Environ 18(2020):100298. (in Press)

Reid WV, Berkes F, Wilbanks T, Capistriano D (2006) Bridging scales and local knowledge in assessments: concepts and applications in ecosystem assessment. Island Press, Washington, DC

Ross A, Pickering K (2002) The politics of reintegrating Australian aboriginal and American Indian indigenous knowledge into resource management: the dynamics of resource appropriation and cultural revival. Hum Ecol 30:87–214

Siahaya ME, Hutauruk TR, Aponnon HSES, Hatulesila JW, Mardhanie AB (2016) Traditional ecological knowledge on shifting cultivation and Forest Management in East Borneo, Indonesia. Int J Biodiv Sci Ecosyst Serv Mgmt 12(1–2):14–23

Sinjela M, Ramcharan R (2005) Protecting traditional knowledge and traditional medicines of indigenous peoples through intellectual property rights: issues, challenges and strategies. Int J Minor Group Rights 12:1–24

Siti ZM, Tahir A, Ida Farah A, Ami Fazlin SM, Sondi S, Azman AH, Maimunah AH, Haniza MA, Siti Haslinda MD, Zulkarnain AK, Zakiah I, Wan Zaleha WC (2009) Use of traditional and complementary medicine in Malaysia: a baseline study. Complement Ther Med 17:292–299

Toy T (2015) Some lesser known facts about *jhum* in Nagaland, Northeast India. In: Cairns MF (ed) Shifting Cuntivation and environmental change: indigenous people, agriculture and Forest conservation. Earthscan, Routledge, London; New York, pp 468–472

Turnhout E, Bloomfield B, Hulme M, Vogel J, Wynne B (2012) Conservation policy: listen to the voices of experience. Nature 488:454–455. https://doi.org/10.1038/488454a

UNDRIP (2007) United nations declaration on the rights of indigenous peoples (UNDRIP). United Nations, London

Wan ZI (ed) (1996) *Kami Bukan Anti-Pembangunan!* [We Are Not Anti- Development!]. Persatuan Sains Sosial, Bangi

Yamyin W, Wittayapak C, Leepreecha P, Mangkhang C (2020) Indigenous rice: the construction of peasant identities base on cultural ecology. Geografia-Malaysian Journal of Society and Space 16(3):14–25

Yin L, Xue D, Wang J (2015) Rubber plantation, swidden agriculture and indigenous knowledge: a case study of Bulang village in Xishuangbanna, China. In: Cairns MF (ed) Shifting Cuntivation and environmental change: indigenous people, agriculture and Forest conservation. Earthscan, Routledge, London; New York, pp 811–825

Tagal System: A Biocultural Conservation Approach in Sabah, Malaysian Borneo

11

Jurry Foo, Dg Nur Fairuzzimah Ag Samara, and Colonius Atang

Abstract

This chapter aims to discuss biocultural conservation via the Tagal system practice among the local community. The Tagal system is a very popular practice in community-based resources management in Sabah, Borneo, Malaysia. This system has been practiced continuously by community members; it has been proven to be an effective approach in conserving and regenerating forest and river ecosystems for a long duration. It is also supported by the strength of cultural principles and customary beliefs, which are highly important factors in human affairs. We also present a description of the concept of biocultural conservation, followed by the implementation of the Tagal system. Data used in this work is based on a case study derived from prior studies. Finally, we offer some suggestions to sustain the Tagal system in the local community.

Keywords

Biocultural conservation · Community-based resources management · Conserving · Regenerating · Culture · Customary · Willingness · Tagal system

J. Foo (✉)
Geography Programme, Faculty of Social Sciences and Humanities, Universiti Malaysia Sabah, Kota Kinabalu, Sabah, Malaysia

Borneo Institute for Indigenous Studies, Universiti Malaysia Sabah, Universiti Malaysia Sabah, Kota Kinabalu, Sabah, Malaysia
e-mail: jurryfm@ums.edu.my

D. N. F. A. Samara · C. Atang
Geography Programme, Faculty of Social Sciences and Humanities, Universiti Malaysia Sabah, Kota Kinabalu, Sabah, Malaysia
e-mail: zimahfairuz@ums.edu.my; colonius@ums.edu.my

S. A. Abdullah et al. (eds.), *Conserving Biocultural Landscapes in Malaysia and Indonesia for Sustainable Development*,
https://doi.org/10.1007/978-981-16-7243-9_11

11.1 Introduction

The conservation of biocultural landscapes is vital to ensure the sustainably of livelihood. Conservation is an action to control the usage of certain resources and appreciate their quality by the current generation (Robert 2009). It is also referred to as the action of taking care of the environment (Cooke 2009). Humans influence the earth system through their activities (Ramzah and Jurry 2009). Human and non-human components are equally important to the biocultural landscape. Hence, to maintain biodiversity, sustaining its cultural aspects is needed (Rosazman 2009). Humans, as a component in the ecosystem, have designed the biocultural landscape via their interests, perceptions as well as their reactions to the surrounding environment. The socio-ecological concept can be used to explain how the community has optimized its local wisdom in conserving the biocultural landscape.

Environmental protection is among the principles included in the Sabah Development Corridor (SDC) blueprint towards the sustainable development of the state, particularly for the eco-tourism, agricultural, and manufacturing sectors, through biodiversity utilization efforts. In Sabah, State of Malaysian Borneo (hereafter refer to as Sabah), a community-based River Tagal system was established for the management of the biosphere system, principally for the riverine ecosystem, which has been proven to be effective in the conservation of the natural and cultural heritage. Although, fundamentally, this system is intended for the conservation of natural heritage in terms of execution, tangible and intangible cultural heritage conservation has been deemed complementary to the success of this system. This work examines and discusses the implementation of biocultural conservation via the river Tagal system in Sabah, Borneo, Malaysia.

11.2 Biocultural Conservation

Biocultural refers to human and non-human aspect in an ecosystem the so-called biocultural landscape. The meaning of cultural landscape has been discussed since the early of the year 1970 as human impact as well as "ecological outcomes" (Williams 1997). The concept of biocultural diversity is related to the variety of the world's cultures as well as the natural environment (Robert 2009). The issues of biocultural diversity focus primarily on participatory development, as well as on social and biological sustainability (Cooke 2009).

In the socio-ecological system (SES), humans play the role of the drivers of ecosystem dynamics and interact with the Earth. This means that human societies are embedded within nature and the surrounding environment. Ecological and social systems need to be examined and managed as integrated SESs; this is related to the degree of capability of the system in terms of self-organization, learning, and adaptation while continuing to develop (Marjolein et al. 2017). A SES consists of human and ecological components that are divided into social and ecological aspects. The social component refers to human activities, including economy, technology, politics, and culture. The ecological component is related to the element

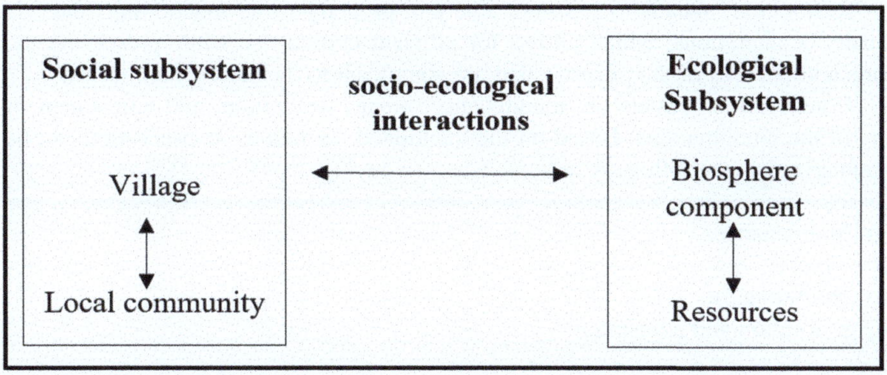

Fig. 11.1 Socio-ecological system approach is applied in utilization of local wisdom to conserve biocultural landscapes in Ranau, Sabah, Malaysia. Source: Adapted and integrated from Santiago et al. (2021) and Carter et al. (2014)

of biosphere, which is understandable in an ecosystem (South American Institute for Resilience and Sustainability Studies 2021).

SES approach was applied to explain the utilization of local wisdom to conserve biocultural landscapes in Ranau (Santiago et al. 2021) (see Fig. 11.1). The redesigned SES concept comprises three components, namely social system, ecological system, and feedback or socio-ecological interactions. The social subsystem consists of villages and local communities (local wisdom), while the ecological subsystem comprises the biosphere component and resources. The dimensions of these subsystems (villages, local communities, biosphere component, and resources) are interrelated and influence each other. A change in one dimension will affect the others through socio-ecological interactions.

Conservation efforts are primarily carried out to halt the further decline, rotting, or outdated state of buildings, monuments, and landmarks via the repair of structures and periodic maintenance (National Heritage Department 2018a, b), as stated in the 2005 National Heritage Act:

"Conservation" means to halt further decline, rotting, or outdated state and provide safe and solid structures, without the expectation of substantial re-development, and entails:

(a) techniques to cease or decelerate the process of decline, rotting, or outdated state of certain details or structures,
(b) the repair of a structure to make it safe, inhabitable, usable, and,
(c) regular maintenance and minor repairs which do not alter or cause further harm to the appearance of the fabric or history of the structure.

Meanwhile, restoration and conservation are defined as:

"Restoration" means to reinstate a certain property to its functioning state via repairs or alterations, which allows for efficient current use while preserving the parts and features of the property that are significant to its historic architecture.

"Conservation" entails the maintenance, repair, restoration, and adjustment or any of the combinations. Based on the mentioned definitions, conservation can be summarized as the following actions:

(a) Maintenance,
(b) repairs,
(c) rebuilding, or,
(d) restoration and adjustment.

The basic principles of conservation are internationally approved by the members of UNESCO and entail actions that require minimal intervention, application of scientific studies, as well as pre-, during-, and post-conservation works (National Heritage Department 2018a). This underlines that cultural heritage conservation does not only entail the ongoing maintenance and protection of heritage materials, but also the attention and intervention of research and development and intensive knowledge retention.

11.3 Biocultural Conservation Via the River Tagal System

Rivers are significant to the livelihood of the people living in Sabah as the source of food supply, especially freshwater fish, and are used for domestic necessities. The Tagal or *Bombon* system is a community-based approach for conserving the biosphere system in Sabah, particularly for restoring the quality of riverine ecosystems and managing the extinction of species, a phenomenon that came to light in the 1960s (Jurry 2011, 2019a; Jephrin et al. 2009). For this purpose, several basic objectives of the *Tagal* system have been established, which are as follows:

1. To protect and conserve the riverine environment, including its ecosystems, habitats, and waters from damage and pollution.
2. To protect and sustain the near-extinct riverine fishery resources.
3. To increase the production of river fish as one of the protein-based food sources for rural communities.
4. To conduct various income-generating activities within the *Tagal* zone such as tourism, for the benefit of the rural communities.

These objectives indicate that the implementation of the *Tagal* system does not only benefit the natural environment, but also the livelihood of the general community living there. This system even shares the same principles with that of cultural heritage conservation. The *Tagal* system was adapted from the forest *Tagal* system that has been traditionally practiced by the Dusun or Kadazandusun community.

The Sabah State Department of Fisheries (DoF) has taken the initiative to investigate and further improve the traditional system and making it more dynamic, universal, and applicable to all aquatic ecosystems. The river *Tagal* system was subsequently implemented on riverine villages in Sabah by the local communities, under the supervision and guidance of the DoF. The DoF also provides assistance in the operations, promotion, research, and preparation of materials for the executors (The Sabah Fisheries Department 1997). This has contributed to the system's recognition as a smart sharing system in the matters of conservation, protection, and specific management practices related to environmental, land, and water preservation, between the local community and the Government of Sabah (Normala et al. 2007). Meanwhile, this smart sharing system emphasizes the action of protection, reviving, and management.

The Tagal system has also been implemented in several aquatic ecosystems in the coastline areas and paddy fields. Although the implementation of the Tagal system is the same for all riverine areas, the management method used by each community is unique, as it is subjected to specific customs and cultures. In this regard, the Tagal communities are entrusted with the responsibility to determine the rules and methods of implementation, according to their respective local customs and beliefs, to ensure that the system is community-friendly and easily accepted by the community members.

The Tagal system is a traditional conservation method that has been trusted for several generations (Normala et al. 2007; Er Ah Choi et al. 2012). Hence, traditional knowledge, values, and practices have been gathered and improved as part of the intangible cultural heritage. This includes the practice of "*sogit*," recognition of authority for the Native Court and Representative of the Native Chief (WKAN), as well as the application of village customs. The Tagal community endeavors to preserve their identity of location as well as traditions and culture. In this respect, the Tagal system helps the local community to conserve both the biological and cultural resources available.

The river and all the biological components in the system are biocultural elements conserved via the river Tagal system. The system serves food supply, particularly protein, and is continuously reliant on fishery activities (Jephrin et al. 2009). Currently, over 554 Tagal sites in Sabah have been managed using this system. Rivers managed under the Tagal system are enforced under the village rules (also known as village customs) as well as under the jurisdiction of the Customs Law and Enactment of Land Fisheries and Aquaculture 2003. Meanwhile, the local communities are authorized to protect and manage the fishery resources using the Tagal system, subject to Sect. 58 of the Native Courts Rules 1995 (PACOS Trust 2008; Jephrin and Moris 2007).

Additionally, the most protected fish species via the Tagal system conservation are those deemed special by the community and often make up their traditional food staple, such as the *Tor* sp. or *Sabah Masheer Tor* spp. from the Cyprinidae family, including the *Tor douronensis* and *Tor tombroides*. These species are known as Pelian, *Garap* or *Pait* by the local community. They have unique features and are expensive and highly sensitive to disruption by their surroundings, whereby they can only survive in clean, flowing water and usually reproduce in the depths of the river.

Species from the Cyprinidae family that are often avoided are the Barob (*Barb sp.*), which are deemed as predators by other fish species.

Other species being conserved under the Tagal system are the Balat (*Phylum Echinodermata*) and lobsters (*Macrobrachium rosenbergii*). However, the objective for the conservation of these species is mainly for meeting market demands. Meanwhile, the conservation and introduction of a new species known as "goldfish," "Japanese carp," or "a-koi" (*Cyprinus carpio*) from the Cyprinidae family is part of the species diversification efforts believed to be a potential eco-tourism product. To ensure the overall well-being of the aquatic life population, plant species that complete the food chain are maintained and conserved, including fruit-producing natural plants at riverbanks such as Pangi (*Pangium edule*), Nunuk (*Ficus sp.),* and Marang Parang (*Seraca sp.*) (Jurry 2011).

Other than biodiversity, some parts of the river under the Tagal system prohibit the extraction of physical natural resources such as rocks, sand, and river water, without the permission of the village authorities. For instance, the Tagal area in Kg. Pinawantai, Ranau, drafted a communal agreement against the extraction of physical resources, as stated in the 2009 Official Letter of the Establishment of the Tagal Committee:

> "..3. Prohibition of the extraction of rocks and sand (for outsiders and contractors unless with the permission of the Tagal Committee. . ."

Tagal Committee, Kg. Pinawantai (2009).

The rationale for the prohibition is primarily to maintain the health of the riverine system and to reserve supply for future needs by the community.

Cultural heritage is highly relevant to environmental protection,especially for biodiversity resources through community involvement and sustainable use of these resources. Intangible cultural heritage is extremely important in the conservation of riverine ecosystems via the Tagal system, especially in matters concerning customs and rules. This is because the link between humans and environmental diversity is highly influenced by beliefs and religious systems which act as the catalyst to harmony (Normala et al. 2007). In fact, for the local communities in Sabah, humans and the environment are culturally and spiritually linked (Felix 2002; Jannie and Colin 2007).

11.3.1 The Role of Village Customs and the Native Courts

The application of village customs and the involvement of the Native Court (MAN) indicate the acceptance and continuation of the Tagal system traditions. MAN is the traditional institution regulating the affairs of the local communities in Sabah, particularly among the Bumiputra ethnics. It was established in the State in the 1880s to ensure the adoption and implementation of the native customs law. In respect of the Tagal system, MAN is responsible for monitoring the adaptation of the customs rules into the system. One example is the endorsement of the by-laws for the management of the Tagal system in Sugut Perai, Beluran by MAN, prior to its actual implementation. This is aside from the management of cases of misconduct around the Tagal areas in instances when they are unsolvable at the village level (Pengurusan SFM SCA Trusan FR 2015).

Meanwhile, customs are rulings that are based on the beliefs, values, and norms of certain communities, which must be observed as a sign of respect to the natural laws of life and of which has become the norm in their daily activities. Customs entail the management of the natural environment by traditional communities and are usually carried out without being formally documented, apart from being orally inherited from one generation to the next, courtesy of the elders or parents in the community (Jannie and Colin 2007; Godfrey 2017). Customs typically form the features of the local justice system in maintaining order and creating harmony in the community, as the customs rules do not only affect individuals, but also the community itself (Felix 2002).

Traditions, however, are rules generally accepted by the society as guidelines in implementing laws, inherited from ancestral practices, and normalized to maintain tradition (National Heritage Department 2018b). In the context of the management of the Tagal system, village customs are community traditions that are implemented to complement the methods of implementation, particularly in the interest of the community members. Among the retained customs is the practice of *sogit* in *monogit*. As village customs are highly specific to the community, the practice is unique from one Tagal area to another.

11.3.2 Sogit

"Sogit" is the symbol of respect. The local communities in Sabah have been taught to respect the natural environment since childhood. They have also been taught to believe that all natural elements, including plants, rivers, the sea, and stones possess supernatural forces, and that earth is inhabited by good and evil spirits which are contactable via a ritual involving the *rinait*, i.e. the dictation of specific spells to ask for permission to use the natural resources found in their surroundings (Minah 2013). In the Tagal system, such rituals are normally carried out during the officiation of events or gazetting of areas for the implementation of the *Tagal* as well as during the opening of *Tagal* areas. It is during these events that the *sogit* ceremony will be carried out.

Sogit is a term in the Dusun language that means "cold," "cooling," or "penalty" and is subject to the context (Low and Solehah 2018; Pg. Ismail et al. 2018). *Sogit* is basically a neutralizing offering. According to Felix (2002), *sogit* is not a form of punishment, but more of an opportunity for the offender to seek forgiveness and subsequently maintain unity. Commonly, *sogit* is carried out through a ritual called *monogit,* or without any ritual in a traditional event based on customs that depend on the situation and purpose. There are two situations that require the *monogit* in relation to the exploration of resources and areas. The first is when permission is needed to explore certain areas or use certain resources, while the second is when problems or catastrophes occur during or after the exploration. In the first situation, *sogit* is imposed to ask for blessings and permission to use certain areas for particular events, especially for exploring natural resources. The *monogit* is then carried out with rituals and sacrifices in the form of animals known as "blood sacrifices." In the second situation, the ritual is carried out when catastrophes, illnesses, or disturbances befall the explorers or users of the resources. Such occurrences are believed to be caused by the *rogon*, an evil spirit whose wrath is incurred due to being disturbed and of whom prior permission was not sought, consequently causing a "hot" situation. Hence, *monogit* is believed to be able to restore or to "cool" the situation. The elders or senior members of the community who are knowledgeable about the customs, or *bobolian* (i.e. witch doctors), will take the lead in these rituals (Jurry 2019a).

Sogit is also a custom-based penalty imposed upon certain parties to restore harmony in the village. A common misconduct that must be penalized is the act of purposely polluting communal water sources, ponds, or rivers (John 2002). The penalty is usually in the form of animal offerings such as buffaloes, pigs, goats, and chickens, in the form of agricultural produce such as paddy, or even in material forms such as the gong, ornaments, weapons, and plates. However, in contemporary times, the s*ogit* is usually in the form of cash deemed as compensation and a symbol of apology by the offender.

11.3.3 Sogit during the Opening of Tagal

The opening of the Tagal constitutes the harvesting of resources yielded from the areas under the Tagal system, which is carried out by all the members of the Tagal community (Jurry 2019b). The opening of the Tagal is carried out after the Tagal authority, after members of the community have discussed and reached an agreement about the Tagal zones, parties that will be involved, and the date of the event. Some Tagal communities organize major events as a way to promote the Tagal system to the public. Today, the opening of the Tagal has become a product of tourism filled with local arts, cultural programs as well as recreational activities. Cultural features retained in the event include the *rinait*, *mitabang,* and *moginakan* (general feast). Additional activities include community sports and fishing competitions. *Mitabang* is when the members of the community work hand-in-hand to prepare equipment to catch fish, to distribute the catch, and to prepare the general feast. Meanwhile, *moginakan* is held as a general feast to symbolize gratitude.

In the practice of traditional conservation, the equipment that will not cause a major impact are allowed during fish and aquatic harvesting activities. Poison to the so-called *tuba* is also prohibited, even if the poison is made out of natural sources. Among the traditional fish harvesting equipment allowed are *Siud*, fishing rods, hooks, spears, nets (Barnard 2002), *sauk*, *rambat* (nets) (Fig. 11.2a), and *bubu* (Minah 2013). *Siud* or *sauk* (Fig. 11.2b) are nets that are tied to a round rattan or metal and usually have a handle that would facilitate the process of fish harvesting. A

(a) (b)

Fig. 11.2 *Rambat* (**a**) and *Siud* (**b**) are among the traditional fish harvesting equipment allowed in the Tagal system

fishing rod is a fishing line tied to a bamboo stick with a fishhook at the end of the line. A spears or *bujak* is a sharp metal, while *bubu* is basket-like equipment placed in areas with streaming water to trap fish.

11.4 Challenges in Practicing the Tagal System

Cultural heritage conservation efforts are often met with various challenges. Among the main issues in the implementation of the Tagal system is the readiness of the community to participate and comply with the rules enforced by the system. This is because the commitment requires a change in lifestyle and access restrictions to the resources for certain periods of time. Other long-standing, unresolved debates are issues concerning ownership, rights, and benefits. The primary challenge is in ensuring that all the parties involved are satisfied with the methods and outputs of the Tagal system and obtain optimum benefits from it. Additionally, another highly debated issue is the enforcement of rules on trespassers, pollution of rivers caused by anthropogenic activities such as farming and agriculture, infrastructure development and logging. Godfrey (2017) listed several issues and challenges in the implementation of the Tagal system, namely:

1. Conflict on riverine usage (including usage of riverine resourcessuch as rocks and sand).
2. Management authority.
3. Overlap in land ownership.
4. Conflict on development and other structures.
5. Rivers that have been under the Tagal system for too long.
6. Ongoing misunderstandings between the community leaders and the villagers.

7. Inactive implementation of the Tagal system.
8. Recognition.

11.5 Conclusion

The Tagal system is a biocultural conservation that is relatively important to sustainable livelihood. In the context of community-based natural resources conservation, the Tagal system is considered the best approach in conserving biocultural landscapes, particularly for the local communities of Sabah. Instead of the willingness of the community practicing the Tagal system to conserve their riverine ecosystem, their beliefs in traditional customs and their respect to their owned culture are the key factors for the success of achieving the goals of sustainable livelihood.

The practice of the Tagal system is highly significant and is in line with the goals of SDG 2030, which, among others, can achieve the overall quality of the environment and address issues of poverty. This is because this system has been proven to have added value to human life, especially as a systematic way of managing life by utilizing the environment in a more organized manner while providing socio-economic opportunities such as income and employment opportunities related to the use of Tagal system as a tourism product and informal learning center.

Based on the issues and challenges listed by the Department of Fisheries, Sabah, regarding the continuous practice of the Tagal system, efforts need to be stepped up by adding initiatives related to formal and informal education that channels knowledge and skills related to the Tagal system. In fact, awareness and application activities on functions and values in conserving and appreciating river ecosystems need to be expanded not only by top-down approaches, but also from the grassroots.

To ensure the effectiveness of this system in achieving its objectives, all stakeholders must play a role in executing their respective functions. Meanwhile, the cultivation of the customs and practices of the community must be continued as the basis for more community-friendly management of resources.

References

Barnard MY (2002) Giliran Penduduk Kampung Malanggang 'anjur pesta ikan'. Harian ekspres (16 March)

Carter NH, Vina A, Hull V, McConell WJ, Axinn W, Ghimire D, Liu J (2014) Coupled human and natural systems approach to wildlife research and conservation. Ecol Soc 19(3):43

Cooke F (2009) Biocultural biodiversity: issues and challengers for research and policy. Paper presented. Seminar on Supporting Biocultural Diversity and Conservation in Sabah. Kota Kinabalu, Sabah. Malaysia. 3rd. December

Er Ah Choi S, Sekvadurai N, Lyndon N, Sheau Tsuey Chong, Adam JH, Mohd Fuad MJ, Habibah A, Hamzah J (2012) The Evolvement of *Tagal* on Ecotourism and Environmental Conservation: A Case Study in Kampong Luanti Baru, Sabah. Adv Nat Appl Sci. 6(1):61–64.

Felix T (2002) Traditional Systems of Indigenous Peoples of Sabah, Malaysia. PACOST Trust, Kota Kinabalu

Godfrey K (2017) The "Tagal System" – A Community-Based Fisheries Management. ASIA Regional Workshop on Satoyama Initiatives in Sabah. https://satoyama-initiative.org/wp-content/uploads/2017/04/Panel-4.-Godfery-Kissey_TAGAL-LE-MERIDIAN-30-MAC-2017xx-rev-1.compressed.pdf. Accessed 10 Oct 2020.

Ismail M, Ismail SZ, Rahman NA (2018) Amalan Sogit Ke Atas Anak Negeri Sabah Beragama Islam Di Mahkamah Anak Negeri: Kajian dari Perspektif Hukum. Jurnal Islam dan Masyarakat Kontemporari. Vol. 19. Penerbit Universiti Sultan Zainal Abidin.

Jannie L, Colin N (2007) Malaaysia. The Changing Status of Indigenous and Statutory Systems on Natural Resource Management. In. Helen Leake. Bridging The Gap. Policies and Practices on Indigenous Peoples' Natural Resources Management in Asia. United Nation Development Programme.

Jephrin ZW, Moris C (2007) Berita Perikanan. Sistem Tagal. Konsep urus sumber ikan sungai secara mampan. http://www.dof.gov.my/html/themes/moa_dof/documents/Berita_Perikanan/2007/bpj63/page10.pdf. Accessed 3 Jan 2011.

Jephrin ZW, Seiichi E, Arthur BS (2009) Toward sustainable community-based fishery resources management: The *Tagal* system of Sabah, Malaysia. Dlm. Siri Ekmaharaj *et. al* (Pnyt.) Uplifting Of the Socioeconomic Condition of The Fisher Via Sustainable Fisheries Management. Fish for the people. Vol. 7. Nov. 2. Southeast Asian Fisheries Development Center.

John A (2002) Penduduk kesal Tindakan individu tertertu racun sungai Sook. Harian Ekspres.:16/3

Jurry F (2011) Sistem *Tagal* dalam industri pelancongan: Manfaat kepada komuniti. GEOGRAFIA Online. Malaysia Journal of Society and Space. 7: 1. Hlm. 14–25.

Jurry F (2019a) Managing the environment of development in Malaysia. The tagal system from the perspective of the local community. Lifeways J. 3(3):52–60

Jurry F (2019b) Pemuliharaan warisan Kebudayaan Melalui sistem Tagal sungai di Sabah. Universiti Malaysia Sabah, In. Borneo dalam Kepelbagaian Perspektif. Kota Kinabalu

Low KO, Solehah I (2018) The spiritual significance of Komburongo in the folk beliefs of the Dusunic peoples of North Borneo. Electronic Jurnal Folklore. 71:179–206. https://www.folklore.ee/folklore/vol71/low_solehah.pdf [1/10/2018].

Marjolein S, Ingrid A, Edwin TP (2017) How to conceptualize and operationalize resilience in socio-ecological systems? Curr Opin Environ Sustain. 28:108–113

Minah S (2013) Kepercayaan dan Amalan Masyarakat Kadazandusun dalam Pemeliharaan Alam Sekitar. Prosiding. Persidangan Intelektual Kebangsaan Malaysia.:631–642

National Heritage Department (2018a) Pemuliharaan warisan.

National Heritage Department (2018b) Akta Warisan Kebangsaan 2005 [Akta 645] http://www.heritage.gov.my/ms/kenali-kami/akta-warisan-kebangsaan-2005. Accessed 27 Sept 2018

PACOS Trust (2008) Securing Indigenous Peoples' Rights in Conservation Reviewing and Promoting Progress in Sabah, Malaysia. Working draft. FPP series on Forest Peoples and Protected Areas. 1 September. Forest Peoples Programme.

Pengurusan SFM SCA Trusan FR (2015) Sistem *Tagal* Sungai Sugut Parai. http://www.forest.sabah.gov.my/sugut/PDF/System%20*Tagal*%20Kg%20Terusan%20Sugut.pdf. [21/9/2018].

Ramzah D, Jurry F (2009) Traditional Knowledge on Climatology and biogeography: A way forward for conservation. Paper presented. Seminar on Supporting Biocultural Diversity and Conservation in Sabah. Kota Kinabalu, Sabah. Malaysia. 3rd. December.

Robert FP (2009) Knowledge of Rafflesia sp. Etlingera sp. And Hibiscus sp., and its Influence on the perception and practices of conservation in Sabah. Paper presented. Seminar on Supporting Biocultural Diversity and Conservation in Sabah. Kota Kinabalu, Sabah. Malaysia. 3rd. December.

Rosazman H (2009) Biocultural Diversity and orang Sungai in Lower Kinabatangan area. A reconcile relationship. Paper presented. Seminar on Supporting Biocultural Diversity and Conservation in Sabah. Kota Kinabalu, Sabah. Malaysia. 3rd. December.

Sabah Fisheries Department (1997) The Tagal Initiatives. http://fishdept.sabah.gov.my/tagal2.asp. [Accessed 1 March 2020.

Santiago ZFH, Vargas J, Grande JM (2021) Integrating socio-ecological information to address human–top predator conflicts: the case of an endangered eagle in the eastern Andes of Columbia. Perspect Ecol Conserv. 19(1):98–107

Williams M (1997) An exceptionally powerful biotic factor. In. Humans components of ecosystem. Springer, New York. pp. 24–39.

Empowerment Human Resources Through the Gender Development Index and Their Impact on Sustainable Development in the Rural Coastal Areas with Dynamic System and Multi-Policy Model

12

Harsuko Riniwati, Nuddin Harahab, Tiwi Nurjanati Utami, Lina Asmara Wati, and Wike

Abstract

Development in the maritime sector is inseparable from human resources, including the issue of gender equality. Out of the five programs to increase the Human Development Index (HDI), including in the maritime sector, human resource development receives the most attention. The issue of gender equality has received serious attention in Indonesia in order to achieve the Sustainable

H. Riniwati (✉)
Socio and Economics of Fisheries, Fisheries and Marine Science Faculty, Brawijaya University, Veteran Malang, East Java, Indonesia

Interdisciplinary Post Graduated, Brawijaya University, Malang, East Java, Indonesia

The resilience of Coastal and Fisheries Villages Research Group, Veteran Malang, East Java, Indonesia
e-mail: riniwatisepk@ub.ac.id

N. Harahab · T. N. Utami · L. A. Wati
Socio and Economics of Fisheries, Fisheries and Marine Science Faculty, Brawijaya University, Veteran Malang, East Java, Indonesia

The resilience of Coastal and Fisheries Villages Research Group, Veteran Malang, East Java, Indonesia
e-mail: marmunnuddin@ub.ac.id; tiwi@ub.ac.id; linaasmara@ub.ac.id

Wike
Interdisciplinary Post Graduated, Brawijaya University, Malang, East Java, Indonesia

Master Program of Women's Studies, Malang, East Java, Indonesia
e-mail: wike_fia@ub.ac.id

S. A. Abdullah et al. (eds.), *Conserving Biocultural Landscapes in Malaysia and Indonesia for Sustainable Development*,
https://doi.org/10.1007/978-981-16-7243-9_12

Development Goals agenda. Furthermore, to measure this equality, the Gender Development Index (GDI) has been used in many studies. This index is formulated as the ratio of the female HDI to the male HDI. The GDI value can be used as an indicator for the empowerment of rural coastal communities. This study aims to i. analyze GDI for the empowerment of rural coastal communities, ii. analyze gender-specific differences in the implementation of development in coastal villages, and iii. Formulate recommendations for increasing the GDI to empower communities in rural coastal-based areas. The data were collected from secondary data sources from the Ministry of Women Empowerment and Child Protection and Central Bureau of Statistics of Indonesia. The multi-policy analysis used the assessment of eight experts on criteria, scenarios, policies, and actions in Sustainable Development in Rural Coastal Areas. The data was collected by an expert based on three dimensions of Human Development Indicators: health, education, and a decent standard of living. The results showed that the HDI of females is lower than that of males. This gap is reflected in the differences in the expected length of schooling, the average length of schooling, and life expectancy. Therefore, a systematic improvement in the quality of development for females is needed. This study recommends improving the quality of human resources, increasing the influence of females representation in parliament, and improving the economic contribution. In addition, this study also suggests yearly periodic monitoring and evaluation of the variables of the length of schooling, the average length of schooling, and life expectancy. The best outcome for reducing the gender gap is by improving all development variables for women. The results of the evaluation of policy-related measures with multi-policy analysis showed that GDI plays the most role in the sustainable development of rural coastal communities.

Keywords

Empowerment · Gender Development Index · Multi-policy · Rural coastal · Sustainable development

12.1 Introduction

Indonesia has very large marine fishery resources and biodiversity. The workable sea is about 5.8 million km, and the number of fishermen is about 94 percent, which is of a small scale. The problems with the development of fishing communities include the very large marine fishery resources and the main national performance indicator, but the condition of small-scale fishing communities is poor. The main performance indicators of the sea fishing sector are gross domestic product growth, fishermen's exchange rate, production volume, production value, average household income of fishermen, and average income (Kusdiantoro et al. 2019). The main performance indicators focus on the economic aspect, and not comprehensive empowerment indicators. Furthermore, emphasizing empowerment only on the aspects of welfare will make fishing communities very dependent on natural resources. Marine fishery resources have the character of being a common property and depending on the season. Therefore, it is necessary to apply different empowerment from the very beginning in such a way that the development of fishing communities will be successful and sustainable.

The development of innovative research and community service is expected to analyze and apply a new approach for empowerment. The implementation of empowerment with an inverted pyramid hierarchy is intended to solve the development problems of the fishing communities. In addition, inverted pyramid hierarchy means that empowerment starts with the indicator with the highest level, namely decision-making, critical awareness of the problems encountered, participation, and access. Furthermore, it is hoped that the welfare of coastal rural communities is achieved by not depending on natural resources, but also through the management of social capital. The low social index of fishing communities includes decision-making, politics, education, health, intrinsic motivation, networks, cooperation, participation, organizational, political, and social trust. These social dimensions are called intangible social capital (Ramadhan et al. 2017; Riniwati 2011; Riniwati et al. 2020; Utami et al. 2020). In order to be measurable, the success of community empowerment is evaluated by using the Gender Development Index.

12.2 Case Study Site: Gender Development and Human Resources Empowerment in a Coastal Rural Landscape

In this study, the analysis of the Human Development Index (HDI) is used to measure the level of human development in an area, including coastal villages. The variables measured include life expectancy, the average length of schooling, expected length of schooling, and economic contribution. The ratio between the HDI of females and the HDI of males is the Gender Development Index (GDI). The result showed that the HDI of females in the coastal rural-based areas is lower than that of the males, and this indicates a gender gap. The results of the research carried out in coastal and non-coastal areas, based on males, showed a difference where inequality was higher in rural coastal areas. To reduce or minimize the level of gender inequality in coastal rural-based areas, a strategy of increasing each variable for females should be used, at least equal to males or striving to increase each variable higher for females.

To strengthen female human resources, this can be achieved not only by increasing the GDI elements but also by increasing the proportion of females in politics and professional decision-makers (managers). To measure the level of human resource empowerment using the Gender Empowerment Measure, the factors that influence the level of empowerment of female human resources are the proportion of females in parliament, decision-making (managers), and income distribution. The ideal proportion between females and males is 50:50. Increasing females empowerment with the male strategies increases the proportion of females in parliament, managers, and income distribution. The level of Gender Empowerment Measures in coastal rural-based areas is also relatively lower than in non-coastal areas.

12.3 HDI, GDI, and Empowerment: Causal Loop Diagram (CLD)

System Dynamics is a mathematical modeling method and technique for conceptually complex problem models. Solving dynamic systems analysis problems begins with drawing a causal loop diagram. The Causal Loop Diagram (CLD) model or causal diagram is a model used in problem solving and prevention by comparing each factor representing other factors. CLD in HDI, GDI, and empowerment are shown in Fig. 12.1. The Human Development Index (HDI) will increase due to the increase in health regional community-based coastal countryside. Health increases due to increased community life expectancy education and purchasing power parity (Tisniwati 2012; Setiawan and Hakim 2013; Dewi 2017; Prastowo 2010; Iskandar 2017). Public education is determined by the expected number of schooling years and length of schooling rate. Several indexes such as the parliamentary index, decision-making index, and increasing purchasing power parity may increase the Gender Empowerment Measure (KKP 2019; BPS 2019a, b; Mirza 2011; Ponce et al. 2014).

A participatory approach or human resource involvement is needed (Cabiati 2015). The involvement of human resources, including females, is needed to increase the decision-making index. Therefore, the proportion of managers, professional human resources, administrative staff, and technicians can improve the decision-making index. The parliamentary index increases because the proportion of representation in males and females also increases (KKP 2019; BPS 2019b; Ennaji 2016). These indexes were adopted to create a Causal Loop Diagram (Fig. 12.2).

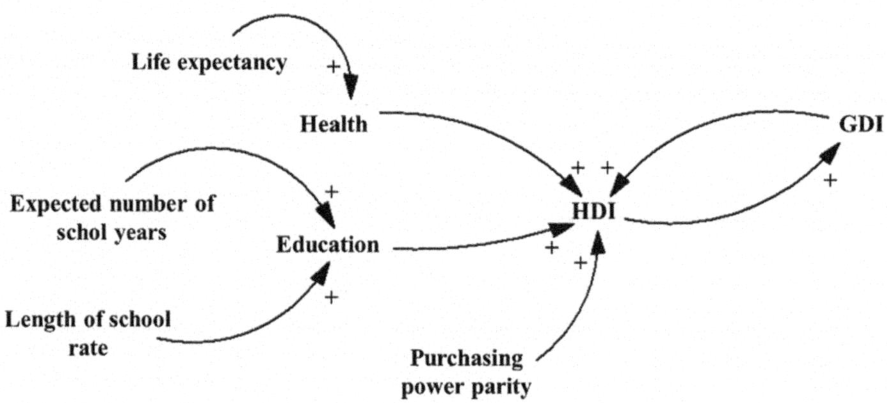

Fig. 12.1 Causal Loop Diagram (CLD) of HDI and GDI

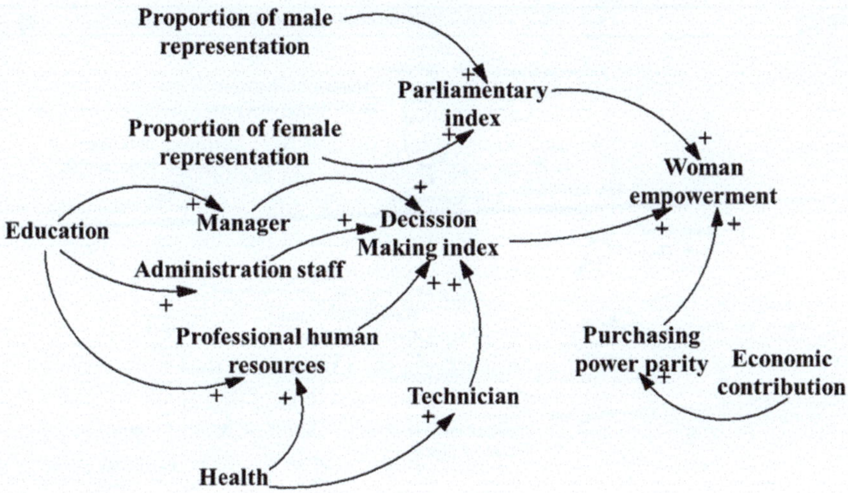

Fig. 12.2 Causal Loop Diagram (CLD) of female empowerment

12.4 HDI and GDI: Stock Flow Diagram

Stock Flow Diagram describes a physical structure, where stock is an accumulation that can increase and decrease, while the flow is a process that leads to an increase or decrease in the stock. The HDI diagram of female, male, and GDI subsystem models is shown in Figs. 12.3, 12.4, and 12.5, respectively. The stock and the flow Diagram are a model that is simulated according to the mathematical formulation. In the GDI simulation system, the main focus is to study the HDI value of males, females, and GDI. There are 3 sub-systems in SFD which are represented in different frameworks, namely the simulation of HDI of females (Fig. 12.3), HDI of males (Fig. 12.4), and GDI (Fig. 12.5). The model formulation is a step taken when compiling stock and flow diagrams in such a way that the model created can be simulated. The formulation is carried out by mathematically putting the relationship between the variables.

The formulation is carried out for all variables. The following is the equation of the GDI simulation model in a coastal rural-based area presented in Table 12.1. A flow chart is created to carry out the simulation. When all variables are connected to numeric units, formulas and functions will be included. The units in the Stella software are called equations. If the variables constant, auxiliary, and level have not been assigned to these units, the program will not be executed. In addition, the units and consistency between variables should be corrected according to the specified scenario for the flow chart template to show the correct quantitative results. The values given in the equations are simulated values, the simulation shows that the HDI value for males is 65.45 and females is 52.43 (Table 12.1).

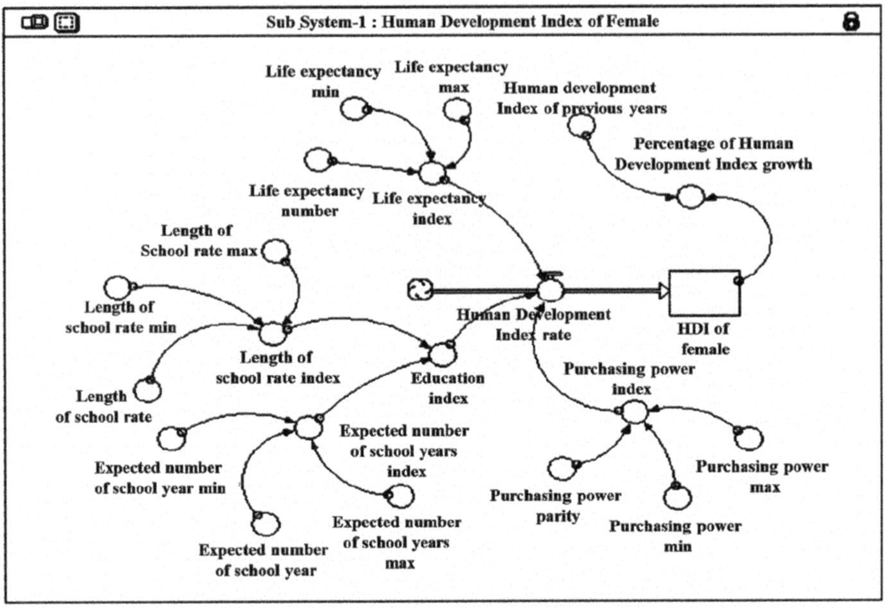

Fig. 12.3 Stock flow diagram of subsystem Human Development Index of female

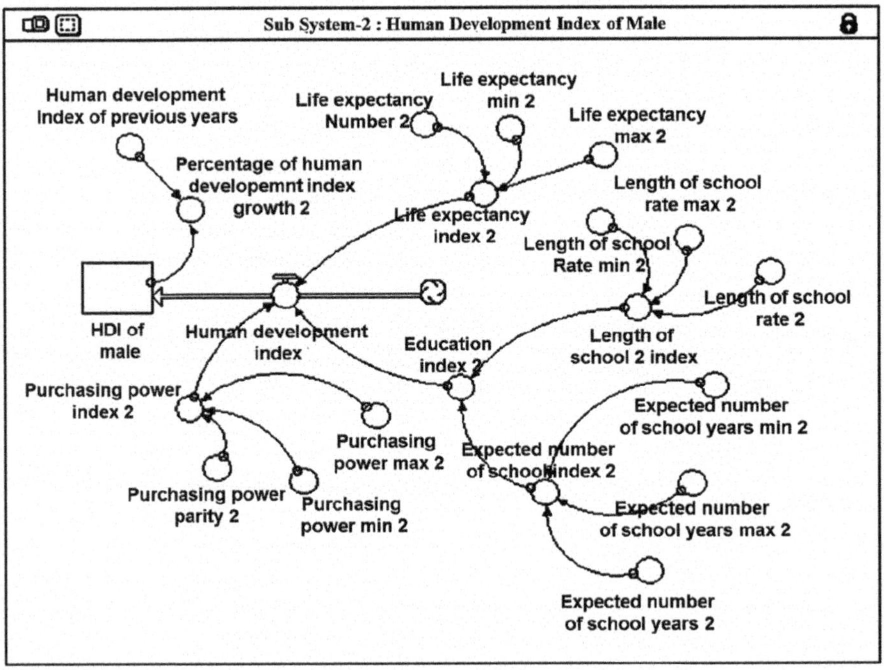

Fig. 12.4 Stock flow diagram (SFD) subsystem Human Development Index of male

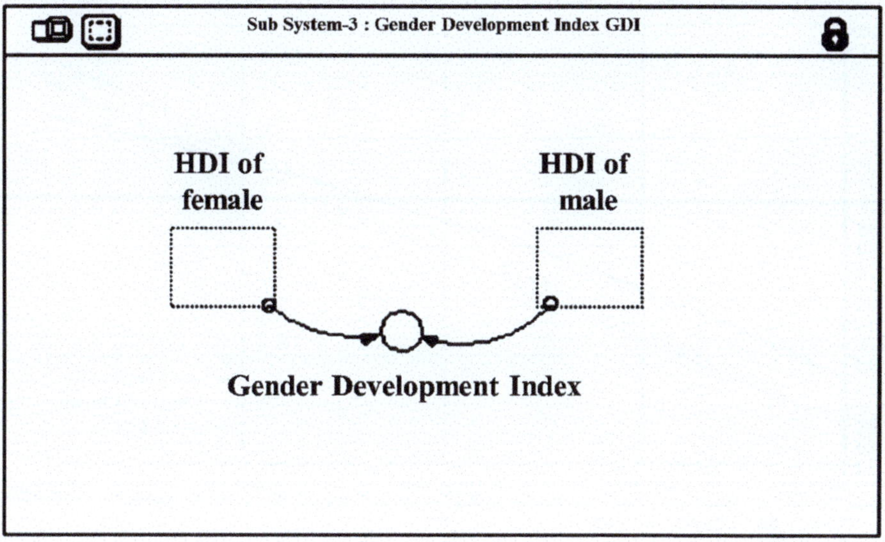

Fig. 12.5 Stock flow diagram (SFD) subsystem Gender Development Index (GDI)

Table 12.1 Summary result of data of Human Development Index (HDI) of male and female

Number	Variable	Male	Female
1	INIT HDI_	65.45	52.43
2	Expected number of school years	10.91	10.54
3	Expected number of school years _max	18	18
4	Expected number of school years _min	0	0
6	human_development_index_previous_year	64.55	51.24
10	life_expectacy_number	63.67	67.22
13	purchasing_power_parity	16.19	13.82
14	purchasing_power_parity_max	17.10	17.10
15	purchasing_power_parity_min	16.51	15.24
16	Length of school rate	7.26	5.7
17	Length of school rate _max	15	15
18	Length of school rate _min	0	0
19	life_expectancy _max	82.5	87.5
20	life_expectancy _min	17.5	22.5

12.5 Data Sources and Dynamic System Model Analysis

With the models and mathematical equations described, simulations of the status of the HDI value for females and males in rural coastal areas (1) can be carried out ten years later. Therefore, it can be used as a reference in formulating gender-based development policies. The results of the analysis with a dynamic system model of

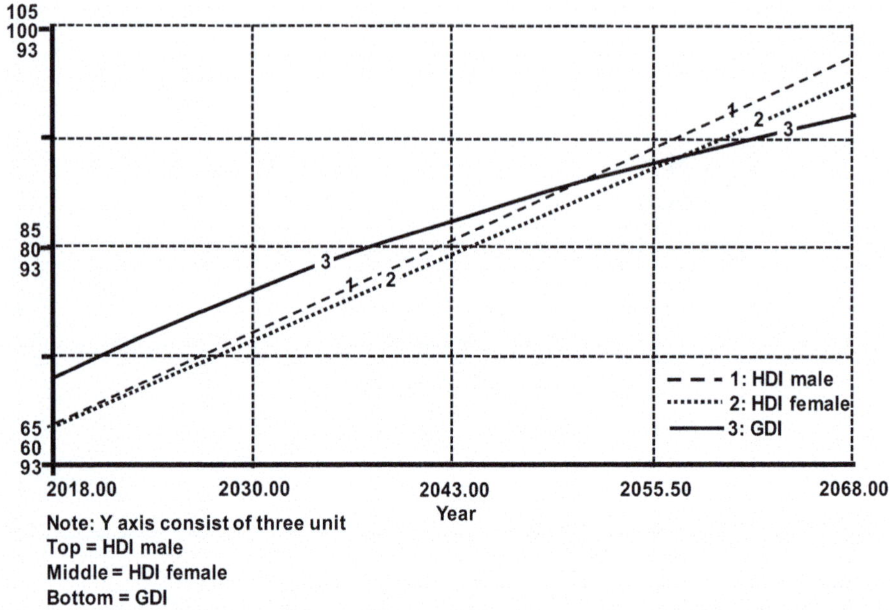

Fig. 12.6 HDI of male, female, and GDI (a real condition in coastal rural areas)

the relationship between HDI of females and males in coastal rural areas (1) are shown in Fig. 12.6. Meanwhile, the current condition of the relationship between HDI of females, males, and GDI in coastal rural areas (2) shows a worse fact, namely a high level of the gender gap. This is shown by the large gap between the HDI of females and males. This situation diminishes the trend of the GDI illustrated by the negative slope GDI graph. The relationship between female, male HDI, and GDI in rural coastal areas (2) is shown in Fig. 12.7.

The interpretation of the simulation model of the dynamic system with the software Stella shows the relationship between the variables in the form of a graphic (Fig. 12.6) as shown in Table 12.2. These data were collected from two study sites in different provinces with large coastal community. The GDI score shows the ratio between the HDI of females and the HDI of males. When the Gender Development Index approaches 100, the gender development will be more balanced or evenly proportionate. However, when far away from 100, the gender development is imbalanced. Furthermore, this should be noted by all policymakers at the central and local government levels as well as related institutions that when the GDI is approaching 100, it can be assumed that the gender gap is getting smaller. However, the ratio of the HDI of females and males, both classified as low can also lead to a GDI value close to 100 and a low gap. Similarly, the HDI of the female and male ratio, which is both high will result in a GDI close to 100 and a small gap. Therefore, the ideal is a high GDI or close to 100, which comes from the ratio of HDI of females and males which are both high.

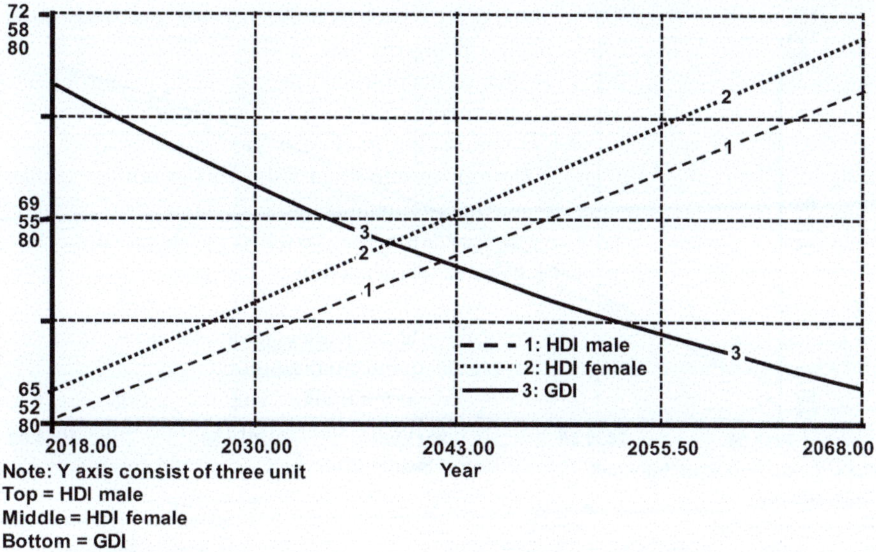

Fig. 12.7 The relationship between HDI of male, HDI of female, and GDI in coastal rural areas

Table 12.2 Trend value of HDI male, female, and GDI

	Coastal Rural Areas (1)			Coastal Rural Areas (2)		
	HDI		GDI = HDI Female/ HDI male	HDI		GDI = HDI Female/ HDI male
Years	Male	Female		Male	Female	
2018	68.36	63.28	92.5688	65.45	52.43	80.10695
2019	69.04	63.91	92.5691	66.1	52.95	80.10694
2020	69.73	64.55	92.5695	66.76	53.48	80.10693
2021	70.41	65.18	92.5698	67.41	54	80.10691
2022	71.09	65.81	92.5702	68.07	54.53	80.10690
2023	71.78	66.45	92.5705	68.72	55.05	80.10689
2024	72.46	67.08	92.5708	69.37	55.57	80.10688
2025	73.15	67.71	92.5711	70.03	56.1	80.10686
2026	73.83	68.35	92.5715	70.68	56.62	80.10685
2027	74.51	68.98	92.5718	71.34	57.15	80.10684

The status of Human Development according to the Kementrian Pemberdayaan Perempuan dan Perlindungan Anak (2012) is classified into 4 criteria. The human development achievements in a region at a given point in time can be divided into four groups. This grouping aims to organize the regions into identical groups in terms of human development. The grouping of human development achievement status is as follows: (1) "very high" group: HDI $\geq$ 80; (2) "high" group: 70 $\leq$ HDI $<$80; (3) "medium" group: 60 $\leq$ HDI $<$70; (4) "low" group: HDI $<$60.

Due to the methodological changes that have taken place, the effects have had an impact on the interpretation of the GDI figures. With the old method, the resulting

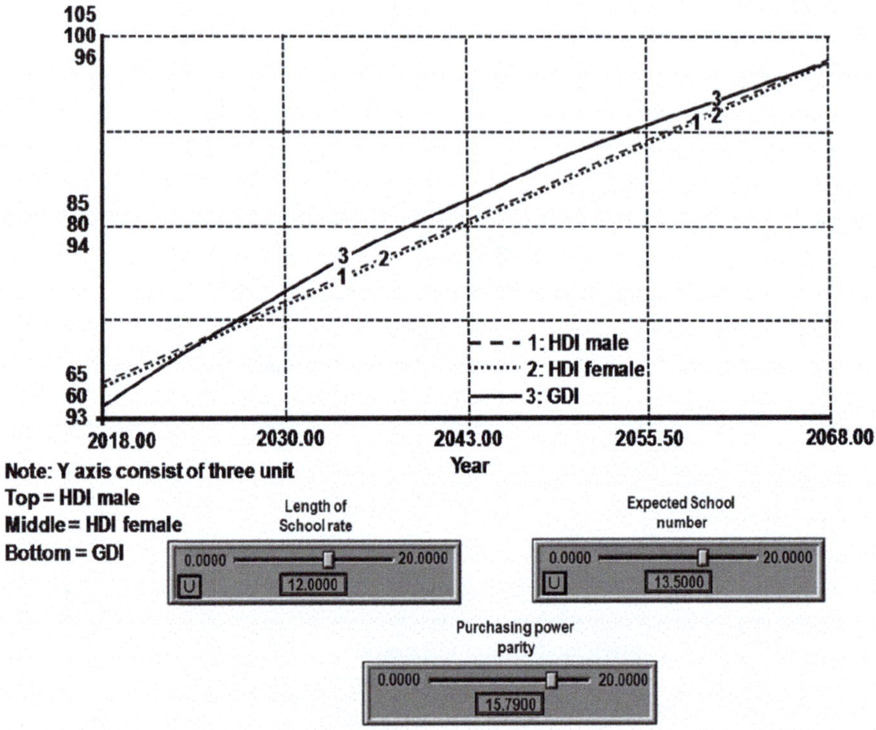

Fig. 12.8 The best scenario with the minimize gender gap

GDI number had to be compared with the HDI figure. The smaller the difference between the GDI score and the HDI number, the smaller the inequality that occurs between males and females. With the new method, the interpretation of the GDI numbers changes, the interpretation of the GDI figures does not have to be compared with the HDI figures. The smaller the distance between the GDI score and the value of 100, the more equitable development is between males and females. However, the greater the distance between the GDI score and the value of 100, the more inequality of development occurs between males and females. The number 100 is used as a benchmark for interpreting the GDI number because this number is the most perfect ratio value.

When considering the trend of the value of female HDI in 2 coastal rural areas in 2018–2027, it is classified as moderate. Meanwhile, the HDI of the male from 2022 to 2027 is classified as high. In order for the GDI value to be determined from the HDI ratio of females to males, which is classified as high, positive measures and an early warning system are necessary. Figure 12.8 shows that there is still an increasing gender gap in rural coastal areas (1) and (2), namely the coarse dash line (1) and the fine dash line (2), which show that the HDI of males and females is getting a wider gap. The impact on IPG has also decreased its trend. The scenario in Fig. 12.8

shows which affirmative action and early warning system programs can be used to control the value of RIB growth in rural coastal areas.

Based on the scenario shown in Fig. 12.8, females' length of schooling rate becomes 12 years, and the expected schooling number of females becomes 13.5 years, the purchasing power parity of females = 16.3 (ln 12,000,000) from the fact 15.79 (ln 7,171,000). The average length of school rate is increased from 5.7 to 12 years and the expected number of years of schooling is increased from 10.54 to 13.5 years in such a way that the gender gap between females and males will be smaller and almost the same and the trend of GDI continues to be positive. As shown in Fig. 12.8, the distance begins to appear, which means that equality between males and females can almost be achieved. The figure shows that the HDI of males and females has a narrower gap, which causes the GDI to increase. The Education for All (EFA) has set the work program to increase the percentage of the world population length of the school of 9 years by 2030. This can be achieved through all forms of education, both formal and informal such as a special program for education.

The influence of the scenario between economic indicators and education has a positive effect on the value of GDI. Related to the increase per capital expenditure, this has become one of the focuses of development. In its presidential speech on October 20, 2019, the Indonesian president focused on 5 development aspects, namely human resource development, infrastructure development, simplification of the bureaucracy, simplification of regulations and policies, and economic transformation from dependence on resources to product orientation that has high added value or industry-based. The target income per capita for the Indonesian people is expected to be IDR 324,000,000 (equal to 21,600 USD) per year or IDR 27,000,000 (equal to 1800 USD) per month with the hope of increasing the human development index for males and females. Based on the results of the simulation with an increase in the per capita income scenario, this will affect the tendency to increase the GDI and achieving gender equality and justice.

COVID-19 has reduced several development indicators, such as the economic indicator. Therefore, local governments need to have a tool in place that can quickly determine which indicators have fallen and what impact they are having on females' HDI and GDI. Due to the impact of the pandemic, the economy is struggling to recover in such a way that the purchasing power parity drops to 15.5, which will have an impact on GDI. However, the scenario length of schooling rate = 9 years, expected years of schooling = 13.5 years but PPP = 15.5, (ln 5,400,000) from initial facts 15.79 (ln 7,171,000) (Fig. 12.9). From Fig. 12.9, it can be seen that there is a decline in economic factors due to the impact of the pandemic. By increasing the factor of education, there is no increase in GDI, although the trend has dropped dramatically.

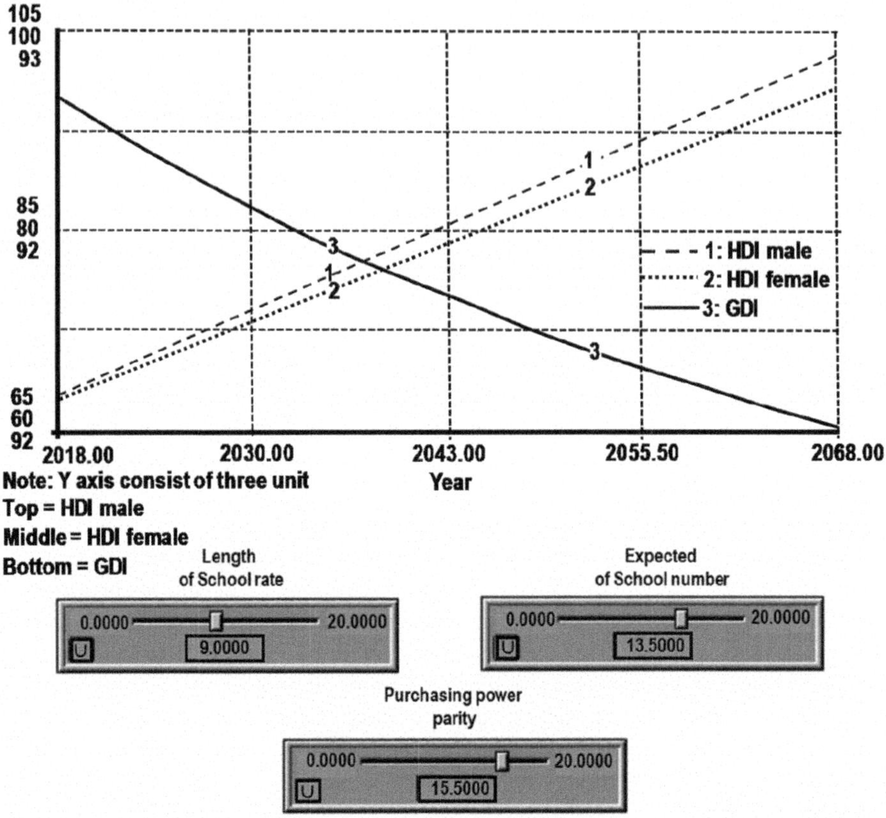

Fig. 12.9 Trend GDI with decreased purchasing power parity of females due to COVID-19 pandemic

12.6 Data Sources and Multi-Policy Analysis

12.6.1 Effect of GDI on Sustainable Development in Coastal Villages

In this study, a multi-policy analysis was used to describe the measures. Using several policies, four components were analyzed including: (a) Criteria: What are the goals and criteria for achieving sustainable coastal village development; (b) Scenarios: Visions, approaches, conditions, or assumptions that may apply or may occur in the future; (c) Policy: the policy or direction to be applied in support of the scenario; and (d) Action: how to implement the policy. The goal of sustainable coastal rural development on the coast is a balance between economy, social environment, accessibility, and cultural preservation (Fauzi 2019). The results of the analysis show that several activities in support of sustainable development include: (i) increased income, (ii) increasing competitiveness, (iii) employment of

Table 12.3 Establishment of goal achievement criteria for "Sustainable Coastal Village Development" in Indonesia

The objectives	Criteria	Long Description of criteria
Economic development	Income	Increased income
	Competitiveness	Increasing competitiveness
Social cohesion improvement	Employment	Absorption of labor
	New business	The growth of new businesses
Environmental protection	Pollution	Reduction of environmental pollution
	Mangrove	Prevention of logging of mangroves
Accessibility	Connectivity	Infrastructure network improvement
Cultural preservation	Culture festival	Cultural festival activities

Sources: modified from Fauzi (2019)

manpower, (iv) new business growth, (v) reduction in environmental pollution, (vi) prevention of mangrove cutting; (vii) infrastructure network improvement; and (viii) cultural festival activities (Table 12.3).

The scenario of "Sustainable Coastal Village Development" in Indonesia has two aspects, namely the demand and supply. Meanwhile, there are 4 aspects of the Sustainable Coastal Village Development policy, namely: (i) multi-function fisheries, (ii) natural resources and cultural tourism, (iii) smart coastal village, and (iv) integrated coastal village. Furthermore, from these scenarios and policies, 7 action plans include (i) the development of fisheries-based tourism in coastal villages, (ii) infrastructure development and strengthening, (iii) developing and improving the quality of fishery products and coastal villages, (iv) development and improvement of information, communication, and technology networks, (v) development of ecotourism based on natural and cultural resources, (vii) increasing the capacity and knowledge of the village community, (vii) improving the formal education of coastal rural communities, and (vii) an increase in the ratio between females HDI and makes HDI (Table 12.4). The data collected from experts in coastal villages can be seen in Table 12.5.

Information from research describing the life profile or portrait of fishing communities in Indonesia indicates that they had a low level of education, income, and life expectancy. This result is consistent with other studies. Typically, the HDI is low in coastal areas because it is characterized by (i) low education, (ii) inadequate housing, (iii) low income, namely between IDR 2,000,000 and 3,000,000 (equal to 133.3–200 USD) per month, (iv) low expenditures ranging from IDR 1,000,000–3,000,000 (equal to 66.7–200 USD), (v) these expenditures cannot meet the social needs such as education costs (education is still neglected), (vi) vulnerable to tidal floods, (vii) traditional fishing technology, (viii) low capital, (ix) welfare of low-income family members, (x) coastal rural development is low, (xi) the life of fishermen is synonymous with poverty, (xii) the welfare of fishermen among other sectors, (xiii) dominant as labor fishermen or crew members (ship crew), (xiv) the number of the female and male population is balanced (Fatmasari

Table 12.4 Evaluation of actions related to criteria in sustainable coastal village development in Indonesia

Short description of the action	Long description of the action
Fishery tourism	Development of fishery-based tourism in coastal villages
Infrastructure	Infrastructure development and strengthening
Product quality	Development and improvement of the quality of fishery products and coastal villages
ICT	Development and enhancement of information, communication, and technology networks
Ecotourism	Development of ecotourism based on local natural resources
Non-formal education	Increase the capacity and knowledge of the village community
Formal education	Increasing the formal education of coastal rural communities
GDI	The increased ratio between the HDI for females and HDI for male

Source: modified from Fauzi (2019)

2014; Marwasta and Kuswaji 2016; Moha et al. 2019a, b; Supriadi et al. 2014; Wasak 2012).

The social aspects that are not affected by the development of fisheries and maritime affairs are indicated by the low social index of the fishing communities. However, capital is grouped into 3, namely physical, human, and social capital. Physical capital is mostly tangible capital, examples include raw materials and infrastructure. Human capital is related to labor, while social capital is classified as intangible. The conditions related to social aspects that are intangible in the fishing communities are still low and even untouched. For example, although the role of intangible social capital lies in social and economic organization, it turns out to be very useful in improving livelihoods and quality of life in the village of Kinabuhutan (Marwasta and Kuswaji 2016). These intangible social capitals are social, organizational, and political participation and trust (Ramadhan et al. 2017).

With the background description above, is there any inappropriate method in building a fishing community? What methods can improve the development success of fishing communities at the national and global levels? It takes a coastal community development strategy with a different approach than the current one, adding the main performance indicators that touch intangible social aspects. The methods that can be used to know the success of human-based development that have taken into account the socio-economic aspects of fishing communities are the human development index, GDI, and GEM. Why should gender development and empowerment be carried out to build a fishing community? This is because (i) the number of females in the fishing community is equal to that of males, (ii) the quality of females is far behind males (gender gap is still high), (iii) social aspects that are intangible for both males and females are still low and have not been touched by fishery development programs. Furthermore, it is also supported by research that showing that social capital serves as a useful frame of mind in successful development and policy formulation (Fathy 2019).

Table 12.5 Evaluation of actions related to criteria of the influence of GDI on the sustainability of coastal rural community issues

Action	Income	Competitiveness	Employment	New business	Pollution	Mangrove	Connectivity	Culture festival	Score
Fish tourism	14	13	12	11	11	12	10	14	2
Infrastructure	13	15	11	13	13	12	16	10	5
Product quality	15	15	13	12	12	10	12	10	3
ICT	14	14	12	14	12	10	15	11	4
Ecotourism	15	13	14	15	13	14	14	16	7
Non-formal education	13	14	13	12	12	10	11	10	1
Formal education	15	14	13	14	12	10	13	13	6
GDI	17	19	19	17	17	17	15	0	8

Note: Grade of position 1 = lowest and 8 = highest

The solution to address the root of the problem is with a different paradigm than the one implemented. Initially, the empowerment process always deviated from welfare and it was no longer followed by increased access, participation, critical awareness of problems, and decision-making, which were the highest level of empowerment. Therefore, an empowerment process is needed that begins with the strengthening of decision-making on resources, which are the most dominant factor influencing empowerment and followed by a critical awareness of problems, participation, access, and ultimately improving welfare. Furthermore, empowerment will have a significant positive effect on intrinsic motivation and performance. Intrinsic motivation is a strong mediator variable in improving performance (Riniwati 2011).

Poverty has a negative and significant effect on HDI. Economic growth has a positive and significant effect on HDI and capital expenditure has a positive and significant effect on HDI. It is suggested that government planning policies aim not only to increase economic growth but also to promote human development, since economic growth alone is not enough to improve the quality of human resources, especially in the areas of education, health, and community income (Mirza 2011).

GDP influences the HDI in the long and short term. The estimate of the ECM model reported that the 2008 economic crisis affected the HDI, while the 1997 crisis and the decentralization of governments did not affect the HDI (Setiawan and Hakim 2013). First, poverty had a significant effect on the human development index in the province of Riau. From the equation, poverty variables showed a coefficient of −0.079, which means that if there is an increase in poverty by 1%, the HDI decreases by 0.079%, meaning every increase in poverty would decrease the percentage of HDI in Riau Province. For the poverty variables obtained, t is greater than the t table, and therefore Ho rejected, while HA accepted. Second, economic growth showed a coefficient of 0.024, which means that if there is a change of economic growth of 1%, there will be changes in the HDI to 0.024%, meaning that any changes in the economic growth will affect the percentage of HDI in Riau Province (Dewi 2017).

The results from the studies of the four variables are likely to influence the HDI in Central Java province, assuming that all other things being equal had a significant positive influence on the HDI from the regency expenditure. The elasticity of the HDI increased to 0.032 due to the increase in expenditure. If the regional spending increased by 1%, the average HDI will increase by 0.032%. The Gini ratio has a significantly negative effect on the HDI. The increased elasticity of the HDI due to a decrease in the Gini ratio is −0.034, if the Gini ratio decreases by 1%, the average HDI increases by about 0.034% The proportion of non-food expenditures has a significant positive influence on the HDI. The elasticity of the HDI increased and the increasing proportion of non-food expenditures amounted to 0.172. If the proportion of non-food expenditure increased by 1%, the average HDI will increase by about 0.172%. The dependency ratio has a significantly negative effect on the HDI. The increased elasticity of HDI in connection with a reduction in the dependency ratio is equal to −0.062. When the dependency ratio decreased by 1%, the average HDI will increase by about 0.062% (Prastowo 2010).

The results showed that the moderation between the human development index through special autonomy fund has a significant negative effect on economic growth.

This indicates that the use of special autonomy funds for the human development index is still high and therefore the future should be allocated according to the needs in the region in such a way that the special autonomy funds can be used effectively and efficiently (Irham Iskandar 2017). Analysis using secondary data with the Multiple Regression Model approach shows that there is one factor influencing the level of poverty and life expectancy, which illustrates the unequal distribution of health services that support population productivity in economic activities to support a better life. Furthermore, with the estimation results of the equation, which results in a decrease in life expectancy by 1%, it has the potential to increase the poverty rate (number of poor people) by 6.9% or 2,245,010 people. Therefore, it is hoped that in the future an effective public policy can be formulated, especially increasing access to basic needs such as education and health (family planning, maternal welfare, basic infrastructure, food, and nutrition) to reduce the poverty rate and not just to decrease the primary school figures but also qualitatively (Tisniwati 2012). Similarly, child empowerment suggested by many authors and professionals should be involved in identifying and maximizing informal resources and formal support (Ponce 2014), need and desire for adopting a participatory approach between parents and service providers (Cabiati 2015), creating progressive social change through females mobilization and participation (Ennaji 2016). Community involvement in a program is essentially empowered marginalized populations and building relationships that can result in future collaboration towards resilience. (Van Auken et al. 2019). A safe source of support will improve welfare and economic empowerment (Stern and Carlson 2019). In addition, entrepreneurship has been suggested as an alternative career model for females to achieve economic empowerment. However, the association with poverty harms the value of HDI, GDI, and GEM, and Rahmah's presentation on gender and culture (Eib and Siegert 2019).

The strategy for increasing GDI is that the percentage increase in the HDI value of females is higher than the percentage increase in HDI of males. The factors that influence the GEM are representation in parliament, managers, professional staff, and economic contribution. The most sensitive is the proportion of the female population, and in parliament, decision-making, and economic contributions.

12.7 The Influence of GDI on Sustainable Development for Rural Coastal Communities

The concept of bioculture is related to the resilience, independence, and sovereignty of rural coastal communities. The impact of GDI on the sustainability of rural coastal communities with a multi-policy analysis shows the most important action to be taken. The results of the analysis are indicated with position 8 (see Table 12.6). This means that GDI has the highest average and the low standard deviation.

While Fig. 12.10 shows that the GDI is the most sensitive measure to improve the achievement of sustainability or the balance between economic–ecological–social dimensions. The accountability of human resources to increase the GDI, therefore, needs to be improved. The methods that can be used to know the success of people-

Table 12.6 Evaluation of actions related to policy of the GDI on the scenario of sustainability of rural coastal communities

Action	Multifunctional Fisheries	Natural Resources and Cultural Tourism	Smart Village	Integrated Village	Mean Value	Standard Deviation	Position
Fish tourism	12.1	12.2	12.2	12.1	12.1	0	2
Infrastructure	12.9	12.8	13	13.1	13	0.1	5
Product quality	12.5	12.4	12.7	12.6	12.5	0.1	3
ICT	12.8	12.7	13	12.9	12.9	0.1	4
Ecotourism	14.2	14.3	14.2	14.2	14.2	0.1	7
Non-formal education	12	11.9	12.1	12	12	0.1	1
Formal education	13.1	13	13.2	13.1	13.1	0.1	6
GDI	17	16.9	17	16.9	16.9	0.1	8

Note: Grade of position 1 = lowest and 8 = highest

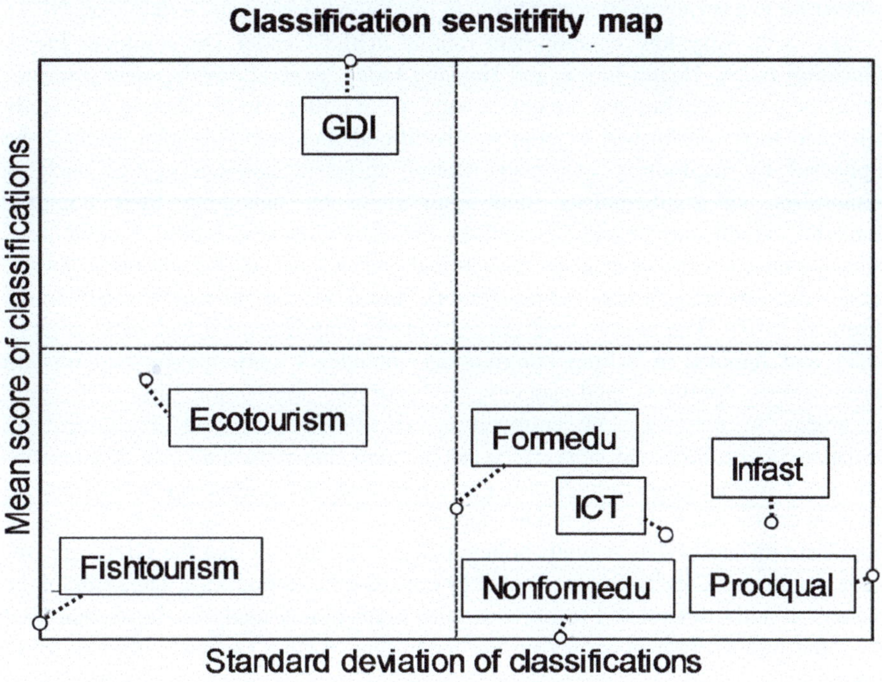

Fig. 12.10 GDI sensitivity map to the sustainability of rural coastal communities

based development that addresses socio-economic and ecological balance and can improve bioculture in rural coastal communities are the HDI, GDI, and GEM. The concept of bioculture is related to resilience, independence, and sovereignty, in this case, the rural coastal communities.

Why should gender development and empowerment be carried out in building fishing communities? This is because (i) the number of females in the fishing community is equal to that of males, (ii) the quality of females is far behind males and the gender gap is still high, and (iii) social aspects that are intangible for both males and females are still low and have not been touched by fishery development programs. Furthermore, it is also supported by research showing that social capital is a useful framework for reflection for successful development and policy formulation (Fathy 2019). The solution for addressing the root of the problem is with a different paradigm than the one implemented. Initially, the empowerment process always deviated from well-being and was followed by increased access, participation, critical awareness of problems, and decision-making, which were the highest level of empowerment. Therefore, an empowerment process is needed that begins with strengthening decision-making on resources, which are the most dominant factors that influence empowerment and are followed by a critical awareness of problems, participation, access, and ultimately will improve welfare. Empowerment will have a significant positive effect on intrinsic motivation and performance, and intrinsic motivation is a strong mediator variable in improving performance (Riniwati 2011). In areas with very low development ratings in the Gender Development Index, Education Index, Health Index, and Housing Index, general development priorities should be identified, but also the nature and type of policies to be taken in each study unit need to be performed to improve livelihood security. This also needs to be supported by facilitating consensus among different partisan groups such as economists, environmentalists, and egalitarians by balancing their common concerns, which can provide guidelines for achieving sustainable development. Subsequently, it can serve as an educational and policy tool to promote a holistic perspective among planners, administrators, and development workers (Singh and Hiremath 2010). The study paradigm shift in the empowerment and motivation of males and females in fishing communities includes i. participation and trust in organizations, politics, and society, ii. intrinsic motivation, iii. Education, iv. health, v. networks, vi. cooperation, vii. Proportions, viii. Parliament, ix managers, x. professional workers, xi. technicians and personnel, xii. Administration, xiii. Creating opportunities, xiv. Creativity, and xv. focus on services. These do not target natural resources or reduce dependence on natural resources by increasing bioculture through GDI. A multi-policy analysis has shown that the GDI improvement program can reach the bioculture of rural coastal communities. The result of the evaluation activities related to policies with multi-policy analysis shows that GDI plays the highest role in achieving the bioculture of rural coastal communities in Indonesia.

Acknowledgments The authors express profound gratitude to the Rector of Brawijaya University, for the support and funding of this research through the project: funding PNBP Fisheries and Marine Science Faculty 2020. Title: Affirmative Action and Early Warning System for Rural Coastal Community Development through the Human Development Index (HDI), Gender Development Index (GDI), and Gender Empowerment Index/Measure (GEM).

References

BPS (2019a) Indeks Pemberdayaan Gender (IDG) [WWW Document]. Indeks Pemberdaya. Gend, IDG. https://www.bps.go.id/subject/40/gender.html#subjekViewTab2

BPS (2019b) Index Pembangunan gender. SIRuSa - Sistem Informasi Rujukan Statistik. Website Badan Pusat Statistik Republik Indonesia. www.bps.go.id

Cabiati E (2015) The need for participative interventions in child protection: perspectives from Nuevo León State. 4:393–420

Dewi N (2017) Pengaruh kemiskinan dan pertumbuhan ekonomi terhadap indeks pembangunan manusia di Provinsi Riau. Jurnal JOM Fekon 4(1):870–882

Eib C, Siegert S (2019) Is female entrepreneurship only empowering for single women. Soc Sci 8: 128

Ennaji M (2016) Article women, gender, and politics in Morocco. Soc Sci 5:75

Fathy R (2019) Modal Sosial: Konsep. Inklusivitas dan Pemberdayaan Masyarakat 6(1):17

Fatmasari D (2014) Analisis Sosial Ekonomi Dan Budaya Masyarakat Pesisir Desa Waruduwur. Kecamatan Mundu, Kabupaten Cirebon

Fauzi A (2019) Teknis Analisis Keberlanjutan. PT. Gramedia Jakarta, Jakarta

Iskandar I (2017) Effect of human development index fund on economic growth through a special autonomy. Jurnal Ekonomi Pembangunan: Kajian Masalah Ekonomi Dan Pembangunan 18(1):50. https://doi.org/10.23917/jep.v18i1.2920

Kementrian Pemberdayaan Perempuan dan Perlindungan Anak (2012) Pembangunan Manusia Berbasis Gender 2012. Kementrian Pemberdayaan Perempuan dan Perlindungan Anak, Jakarta

KKP (2019) Laporan Kinerja Kementerian Kelautan dan Perikanan tahun:2019

Kusdiantoro K, Achmad F, Sugeng HW, Bambang J (2019) Perikanan Tangkap Di Indonesia: Potret Dan Tantangan Keberlanjutannya. Jurnal Sosial Ekonomi Kelautan dan Perikanan 14(2):145. https://doi.org/10.15578/jsekp.v14i2.8056

Marwasta D, Kuswaji DP (2016) Analisis Karakteristik Permukiman Desa-Desa Pesisir di Kabupaten Kulonprogo. Forum Geografi 21(1). https://doi.org/10.23917/forgeo.v21i1.1819

Mirza DS (2011) Pengaruh Kemiskinan, Pertumbuhan Ekonomi, Dan Belanja Modal Terhadap IPM Jawa Tengah. JEJAK 4(2):102–113

Moha I, Sondakh SJ, Kotambunan OV (2019a) Kondisi Sosial Ekonomi Keluarga Nelayan Dan Pengaruhnya Terhadap Tingkat Pendidikan Anak Di Desa Bulawan Induk Kecamatan Kotabunan Kabupaten Bolaang Mogondow Timur. http://ejournal.unsrat.ac.id/index.php/akulturasi Vol. 7 No. 1 (April 2019). ISSN. 2337–4195 / e-ISSN: 2685–4759

Moha I, Sondakh SJ, Kotambunan OV (2019b) Kondisi Sosial Ekonomi Keluarga Nelayan Dan Pengaruhnya Terhadap Tingkat Pendidikan Anak Di Desa Bulawan Induk Kecamatan Kotabunan Kabupaten Bolaang Mogondow Timur. Akulturasi J Ilm Agrobisnis Perikan 7: 1067. https://doi.org/10.35800/akulturasi.7.1.2019.24397

Prastowo A (2010) Menguasai Teknik-Teknik Koleksi Data Penelitian. Kualitatif. Diva Press, Yogyakarta

Ramadhan A, Christina Y, Sonny K (2017) Indeks Sosial Ekonomi Rumah Tangga Nelayan Indonesia. Jurnal Sosial Ekonomi Kelautan dan Perikanan 12(2):235. https://doi.org/10.15578/jsekp.v12i2.6497

Riniwati H (2011) Mendongkrak Motivasi Dan Kinerja : Pendekatan Pemberdayaan SDM, 1st edn. UB Press, Malang, Indonesia

Riniwati H, et al. (2020). Afirmative Action Dan Early Warning System Pembangunan Masyarakat Pedesaan Pesisir Melalui Indeks Pembangunan Manusia (IPM), Indeks Pembangunan Gender (IPG) Dan Indeks Pemberdayaan Gender (IDG)

Setiawan MB, Hakim A (2013) Indeks Pembangunan Manusia Indonesia. Jurnal Economia 9(1)

Singh P, Hiremath BN (2010) Sustainable livelihood security index in a developing country: a tool for development planning. Ecol Indic 10:442–451. https://doi.org/10.1016/j.ecolind.2009.07.015

Stern E, Carlson K (2019) Indashyikirwa Women's safe spaces: informal response for survivors of IPV within a Rwandan prevention Programme. Soc Sci 8:76. https://doi.org/10.3390/socsci8030076

Supriadi D, Agus P, Kasful M (2014) Analisis Potensi Dan Arah Pengembangan Desa Pesisir Di Kabupaten Batu Bara 17(3):14

Tisniwati B (2012) Analisis faktor-faktor yang mempengarui tingkat kemiskinan di Indonesia Jurnal Ekonomi Pembangunan 10(1):33–46

Utami T, Harsuko R, Nuddin H, Lina A (2020) Community participation in development of the seaweed processing industry (study in Madura Island, East Java Province). in Proceedings of the Proceedings of the 13th International Interdisciplinary Studies Seminar, IISS 2019, 30–-31 October 2019, Malang, Indonesia: EAI

Van Auken P, Hauser O, Wolover DJ, Lerch B, Jackson K, Son J, Reed S, Crowley C (2019) Working through uncertainty: the perils and potential of community-engaged research on refugee resettlement. Soc Sci 8:73. https://doi.org/10.3390/socsci8030073

Wasak M (2012) Keadaan Sosial-Ekonomi Masyarakat Nelayan di Esa Kinabuhutan Kecamatan Likupang Barat Kabupaten Minahasa Utara Sulawesi Utara 1:5

Urban Green Space Planning and Management for Biocultural Diversity in Jakarta, Indonesia

13

Jae-Eun Kim

Abstract

More than half of the world's population lives in urban areas, and there is increasing recognition of the importance of urban green spaces in relation to human well-being as cities grow larger. Jakarta, the capital of Indonesia, faces various problems due to rapid urbanization as it is bringing along increasing levels of air pollution and harming the quality of life of residents. The Indonesian government plans to reduce the problems by gradually increasing the area of urban green space through urban landscape planning and management. Like many other urban areas located in tropical regions, Jakarta exhibits a relatively high biodiversity due to its traditional land use patterns. This feature should be utilized in the planning and management of its urban green spaces. In addition, both ecological and cultural aspects of urban and sub-urban areas require attention and should inform the decision-making. Urban green spaces need planning that fits the local characteristics of urban landscapes, both ecologically and culturally. In other words, attention for biocultural diversity will result in more effective landscape planning and management systems.

Keywords

Biocultural Diversity · Landscape Planning · Land Use Pattern · Urban Environment · Urban Green Space · Traditional Land Use · Well-being

J.-E. Kim (✉)
Institution for Marine and Island Cultures, Mokpo National University, Mokpo, Jeollanamdo, Republic of Korea

S. A. Abdullah et al. (eds.), *Conserving Biocultural Landscapes in Malaysia and Indonesia for Sustainable Development*,
https://doi.org/10.1007/978-981-16-7243-9_13

13.1 Introduction

In general, cities exist because of mainly economic and social reasons and urbanization makes many of them grow larger and larger over time. Modern cities have become the centers of politics, economy, and society, and over 50 percent of the world's population now lives in urban areas (WHO-Europe 2017). In the past, upon the emergence of the first cities, the conditions for food cultivation were key in determining which locations were best suited because the community needed to be sustained. For example, early stage cities such as in Egypt and China were located near rivers as these ensured water resources and fertile land (Kaplan and Holloway 2014).

Modern cities are spaces with social and cultural components connected to an economy that are accustomed to a high population density. However, in terms of their operation as a system or ecology (Forman 2014), they are also confronted with various problems such as poor air quality, traffic congestions, issues related to waste management, and substandard housing. Although such problems occur in cities all around the world, typically they are more severe in developing countries. For example, Jakarta, the capital of Indonesia, is facing serious problems such as increasing population density, flooding, and air pollution (Arifin and Nakagoshi 2011; Kusuma et al. 2019; Setiowati et al. 2018), which are problems that directly affect its residents (Kusuma et al. 2019, Setiowati et al. 2018). Increasing demands for improved health, well-being, and quality of life are feeding the desire to improve the sustainability of urban environments. In this, the planning and management of urban green space (UGS) is widely recognized as being of key importance (Van Bueren et al. 2014; Austin 2014; WHO-Europe 2017). Also in the case of Jakarta, UGS has been emphasized to improve residents' well-being and quality of life (Setiowati et al. 2018).

Recently, the planning and management of UGS has been discussed, not only in the sense of simply creating more UGS, but also in respect to increasing biocultural diversity. It is widely agreed upon that UGS contributes to better air quality and more esthetically pleasing spaces. In tropical and subtropical regions, it is fairly common to find high biodiversity in UGS (Mabuhay and Isagi 2005). In addition, these regions are also well known for their cultural diversity. In particular, studies have shown that tropical regions in Asia exhibit both linguistic diversity and biodiversity (Maffi 2005; Maffi and Woodley 2010). The combined concept of biocultural diversity was brought to the attention of policymakers around the world when it was discussed at the 1992 Convention on Biological Diversity. In the United Nations Environment Programme, published in 1999, biodiversity is stated as being related to the culture and spiritual situation of humans. In other words, biodiversity and cultural diversity are tightly linked to each other. This is recognized as a very important factor in discussing sustainable landscapes, both urban and non-urban.

This chapter focuses on investigating the current state of the urban environment in Jakarta and discusses strategies for UGS planning aimed at improving the urban landscape and the quality of life of residents. Moreover, it discusses how biocultural diversity forms an appropriate guideline in creating UGS that makes the urban landscape more sustainable and improves the quality of life.

13.2 Status of Jakarta

In Southeast Asia, Jakarta is one of the urban regions that has seen a rapid increase in population (Fig. 13.1). In 1950, the population stood at approximately 1.5 million and, in the following decade, it grew at an annual rate of over 6%. During the 1960s, 1970s, and 1980s, the population more than doubled with an annual growth rate of 3–4%, bringing the population in 1990 to a total of 8.2 million. After a near standstill in the 1990s due to suburbanization and the 1997 Asian financial crisis (Rukmana et al. 2019), the population continued to grow again at a moderate but steady pace. In the 2000s and 2010s, the annual population growth rate was around 1%, bringing the population of Jakarta to a total of around 10.8 million in 2020.

With its population increasing, Jakarta's land use patterns are undergoing many changes as various urban functions are created or expanded, such as functions to support daily needs. According to Maheng et al. (2021), built-up areas in Jakarta increased by 44.3% from 1995 to 2014, while decreases occurred in land used for cropland (32.27%), grassland (36.11%), and forest (58.05%). According to Setiowati et al. (2018), the area used as UGS in Jakarta decreased by 23% from 1983 to 2013.

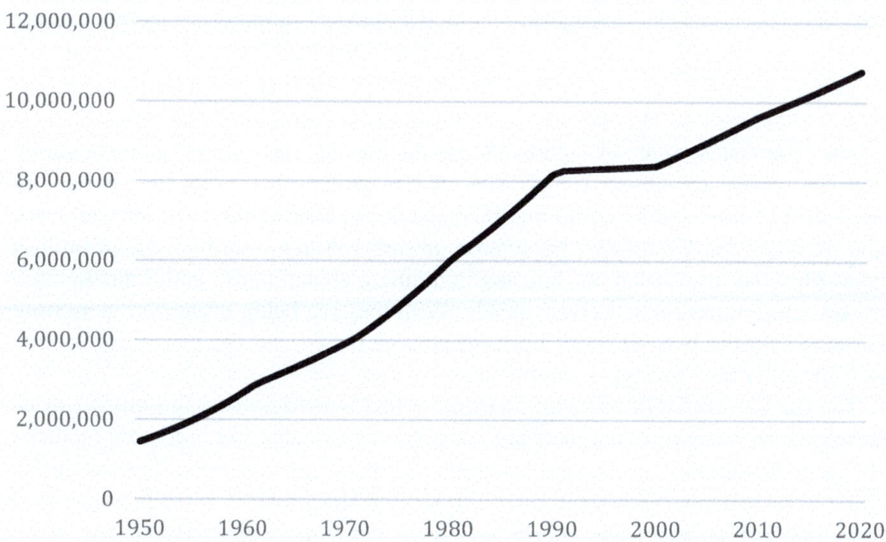

Fig. 13.1 Human population history of Jakarta (data: United Nations-World Population Prospects, https://population.un.org/wpp/)

Table 13.1 Air Quality Index (AQI) levels, related concentrations of fine particulate matter ($PM_{2.5}$), and their health implications (data: EPA (2018))

Air Quality Index (AQI) Values	$PM_{2.5}$ (μg/m^3)	Levels of Health Concern	Health Implications
0–50	0.0–12.0	Good	Air quality is considered satisfactory, and air pollution poses little or no risk
51–100	12.1–35.4	Moderate	Unusually sensitive people should consider reducing prolonged or heavy outdoor exertion
101–150	35.5–55.4	Unhealthy for sensitive groups	The following groups should reduce prolonged or heavy outdoor exertion: • People with lung disease, such as asthma • Children and older adults • People who are active outdoors
151–200	55.5–150.4	Unhealthy	The following groups should avoid prolonged or heavy outdoor exertion: • People with lung disease, such as asthma • Children and older adults • People who are active outdoors Everyone else should limit prolonged outdoor exertion
201–300	150.5–250.4	Very unhealthy	The following groups should avoid all outdoor exertion: • People with lung disease, such as asthma • Children and older adults • People who are active outdoors Everyone else should limit outdoor exertion
301–500	250.5 and over	Hazardous	Health alert: Everyone may experience more serious health effects

The urban changes are also partly due to the fact that the role of Jakarta as capital city has further strengthened. This role creates demand for various economic and political functions and an additional increase in population, which in turn all cause land use changes. Especially, population growth has led to an increase in built-up areas for land uses such as housing, commercial activities, and transportation infrastructure in response to basic needs. With the area being urbanized increasing, automatically the proportion of built-up area increases and the proportion of area used for other land use types diminishes.

The decrease in UGS in Jakarta has had a bad effect on the quality of the urban environment, along with the increasing number of vehicles and industrial facilities. The deterioration of the urban environment in Jakarta has begun to affect the quality of life of its residents (Setiowati et al. 2018). A major problem is the increase in air pollution, such as increasing concentrations of particulate matter (PM). The World Health Organization (WHO) has classified air pollution as the biggest environmental risk to human health, linking it to both mortality and morbidity (WHO-Europe 2017). Table 13.1 shows the health concern raised by different levels of air quality, as defined by the United States Environmental Protection Agency (EPA).

Table 13.2 Average air quality level in Jakarta by Month (Data: 2019 World Air Quality Report, https://www.greenpeace.org/static/planet4-thailand-stateless/2020/02/91ab34b8-2019-world-air-report.pdf)

Month	Monthly average (μg/m^3)	WHO Standard Risk
January	24.2	Moderate
February	34.5	Moderate
March	31.2	Moderate
April	46.2	Unhealthy for sensitive groups
May	58.3	Unhealthy
June	67.2	Unhealthy
July	63.4	Unhealthy
August	53.5	Unhealthy for sensitive groups
September	57.1	Unhealthy
October	60.4	Unhealthy
November	53.3	Unhealthy for sensitive groups
December	43.2	Unhealthy for sensitive groups

Table 13.1 shows the different levels of air quality in terms of concentrations of fine particulate matter (i.e., particles with a diameter less than 2.5 micrometers) and what it means for residents to be exposed to these levels. Fine particles have a strong adverse effect on human health as they can reach deep into the lungs when inhaled (Setiowati et al. 2018; WHO-Europe 2017). Climatic and seasonal conditions also play a role in air pollution. In Jakarta, the air quality has shown to be more favorable when there is a climatic trend of increasing rainfall (Kusumaningtyas et al. 2018) and towards the end of the rainy season in the months of January to March (Table 13.2).

13.3 The Importance and Value of UGS

UGS influences the urban landscape in various ways. It improves the quality of human life by helping cities to adapt to climate change (Douglas 2011) and creating a visually more pleasing urban environment (Bertram and Rehdanz 2015; Tzoulas et al. 2007). Moreover, UGS contributes to the quality of life in terms of counteracting bad urban environmental conditions such as air pollution, urban heat islands, and noise. Increasing contact with nature also brings benefit to people in a psychological sense (Bertram and Rehdanz 2015; Vierikko et al. 2016).

Investments in UGS tend to increase when economic development advances and society stabilizes (Shuvo et al. 2020). In other words, urban areas that are more economically developed and more socially stable experience an increased demand for nature. In case of cities in developing countries, like Jakarta, UGS is typically insufficient in both quantity and quality (Kim 2012; Manan 2016; Setiowati et al. 2018; Ramdhoni et al. 2016). Recently, the Indonesian government has recognized the decrease in UGS due to development and set up a plan to increase the area of UGS step by step (Setiowati et al. 2018). By defining it a priority to increase the proportion of UGS in Jakarta, the government acknowledges the importance of UGS in the urban landscape. However, the quality of UGS is equally important, as it has the potential to affect human health and well-being directly (e.g., Carrus et al. 2015; Fuller et al. 2007; Manan 2015; Zhang et al. 2017). The success of implementation also depends on how urban areas are planned as a whole and to what extent UGS is made an integral part of it (WHO-Europe 2017; Knobel et al. 2021; Lian and Sodhi 2004). Moreover, the level of consistency that is put into the maintenance of UGS will determine whether its potential is fully utilized in the longer term (Ottitsch and Krott 2005).

The UGS management of Jakarta is handled by the government's Park Services, Forest Services, and Agricultural Services. The major problems of UGS in Jakarta are considered to be the reduction of UGS due to development, the lack of proper management, and the lack of awareness among local residents regarding the benefits of UGS (Manan 2016). UGS should be planned in consideration of its structure and function along with the expanding urban landscape. In the strictest sense of the word, UGSs in urban landscapes are not natural landscape elements, as they are made or shaped by humans and not nature. They are landscape elements created with a purpose, and this should be taken into account when planning UGS (Elands and Van Koppen 2012). By using landscape indices to evaluate the spatial patterns, it is possible to predict and assess the landscape function of different UGS scenarios and use this as a basis for decision-making in UGS planning (Turner et al. 2001). From an ecological perspective, it is recommended to create areas of UGS that are large and connected instead of small and isolated, while curved and irregular boundaries are preferred over straight and sharp-angled ones because they perform better in terms of species richness (Forman 1995; Turner et al. 2001; McGarigal and Marks 1995). These aspects are fundamental to establish a spatial pattern of UGS that is more natural and less fragmented and therefore more sustainable and easier to maintain and manage.

In addition to improving the quantity and quality of UGS through planning and implementation, it is also important to consider and define management strategies in advance as well as to set up campaigns to raise awareness of the necessity and value of UGS among residents in the community. The quality of UGS could not be adequately maintained without such efforts. Furthermore, to ensure the sustainability of UGS it is essential to take into account local characteristics, such as the influence of climate and also cultures in terms of how local residents tend to use and perceive nature (Agnoletti and Rotherham 2015; Elands et al. 2015; Stålhammar and Brink 2020). In general, when UGS is accessible, usable, and attractive to local residents, it can provide opportunities for better physical and mental health (e.g., De Vries et al. 2013, James et al. 2015, Maas et al. 2006, Nutsford et al. 2013, Van den Berg et al. 2015).

13.4 Sustainable UGS

For a large number of people who live in cities nowadays, interactions with the natural world are limited to the UGS they have available around them (Dallimer et al. 2014). Especially in developing countries, this is in great contrast with people who live in non-urban areas, such as rural or woodland areas, where people live closely with nature as their livelihood largely depends on it. The reality is that for people who have spent most or all of their lives in urban areas, it is very difficult to know how to live with nature. According to Elands et al. (2015), the concept of biocultural diversity in cities is that in cities, where high urbanization and artificial landscape dominate, biocultural diversity has a creative concept rather than a conservation meaning. Therefore, it seems desirable to consider biocultural diversity in urban areas in a creative sense rather than in the context of conservation of existing organisms and cultures.

Biocultural diversity does not mean that biodiversity promotes or reduces cultural diversity. As in the case of rural landscapes, proper human disturbances (in other words, cultural approaches) affect the urban environment and progress in the direction of increasing species diversity. In other words, biocultural diversity is not a one-way relationship in which nature affects humans as described in ecosystem services (Kim 2019). Instead, it can positively exert influence in both directions, depending on the degree of human disturbances (Stålhammar and Brink 2020). When discussing the concept of the relationship between nature and humans, the viewpoint of biocultural diversity is most appropriately seen in terms of the interactions between nature and humans. In particular, biocultural diversity can strengthen UGS infrastructure planning and landscape management (Vierikko et al. 2016). This potential is due to the fact that it offers a way to match the supply of UGS, both quantitatively and qualitatively, to the specific needs and desires of local residents, which in turn can make sure the UGS is appreciated, well-utilized, and cared-for (WHO-Europe 2017). However, despite the fact that biocultural diversity is a very useful strategy for sustainable development and management of UGS, there have been no political attempts to apply it yet (Elands et al. 2015; Agnoletti and Rotherham 2015).

Traditional rural landscapes are known to have relatively high biocultural diversity due to the existence of proper interactions between nature and humans (Kim and Hong 2016; Elands et al. 2015). Land use and management in these rural landscapes is based on biocultural diversity and is also a resource for biocultural diversity.

Biocultural diversity in urban landscapes should be considered in accordance with the complex and diverse environment formed by the city (Stålhammar and Brink 2020). In other words, especially in urban areas with relatively high biocultural diversity, the planning of UGS needs to be integrated into the city's overall planning, linking it to all other facets at play in the city, including its cultural background. A city like Jakarta, which is known for its rich biocultural diversity, may consider developing a UGS infrastructure with reference to its traditional sub-urban land use pattern and habitat. Through planning efforts that account for the cultural identity of the city and its surrounding areas, UGS can serve as an

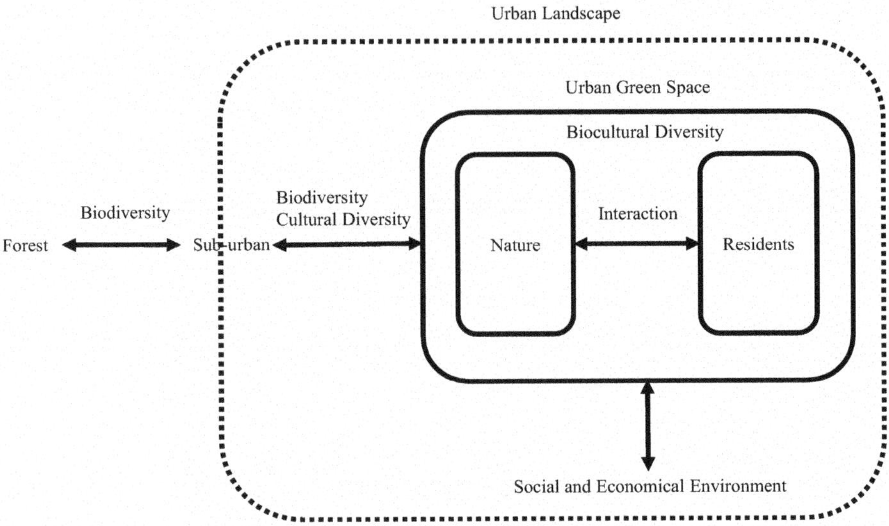

Fig. 13.2 Schematic diagram of sustainable landscape planning and management through biocultural diversity of UGS

instrument to positively affect both the lives of local residents and the sustainability of the whole city (see Fig. 13.2).

13.5 Conclusion

It is important to increase the area of UGS in cities such as Jakarta where the absolute amount of green space is insufficient and proportionally decreasing under the pressure of urbanization. With respect to UGS planning, consideration of the scales of habitats and the connections between different green spaces will be needed. Moreover, using the concept of biocultural diversity in UGS planning will do justice to both nature and humans, and as such it can be a driver of sustainability. Creating UGS based on biocultural diversity can lead to more effective management and more engaged local residents.

UGS can be a means to improve biocultural diversity through landscape planning in which traditional land use patterns of the surrounding area are used as a reference. Traditional land use can enhance natural value and modern management techniques in landscape (UNESCO 2019). As such, it will be possible to develop sustainable UGS by creating a structure that establishes a virtuous cycle in which the UGS promotes the health, well-being, and quality of life of residents, and in which reciprocally the residents are encouraged to make sustainable use of the UGS as it fits their cultural needs and desires.

Acknowledgments This work was supported by the Ministry of Education of the Republic of Korea and the National Research Foundation of Korea (NRF-2020S1A6A3A01109908).

References

Agnoletti M, Rotherham ID (2015) Landscape and biocultural diversity. Biodiv Conserv 24:3155–3165

Arifin HS, Nakagoshi N (2011) Landscape ecology and urban biodiversity in tropical Indonesian cities. Landsc Ecol Eng 7:33–43. https://doi.org/10.1007/s11355-010-0145-9

Austin G (2014) Green infrastructure for landscape planning: integrating human and natural systems. Routledge, London, UK

Bertram C, Rehdanz K (2015) The role of urban green space for human Well-being. Ecol Econ 120: 139–152

Carrus G, Scopelliti M, Lafortezza R, Colangelo G, Ferrini F, Salbitano F, Agrimi M, Portoghesi L, Semenzato P, Sanesi G (2015) Go greener, feel better? The positive effects of biodiversity on the Well-being of individuals visiting urban and peri-urban green area. Landsc Urban Plan 134: 221–228

Dallimer M, Davies ZG, Irvine KN, Maltby L, Warren PH, Gaston KJ, Armsworth PR (2014) What personal and environmental factors determine frequency of urban greenspace use? Int J Environ Res Public Health 11(8):7977–7992

Douglas I (2011) The role of green infrastructure in adapting cities to climate change. In: Douglas I, Goode D, Houck M, Wang R (eds) The Routledge handbook of urban ecology, Routledge London, UK, pp 583–588

Elands BHM, Van Koppen DSA (2012) Biocultural diversity in the Netherlands: from ecologically noble savages towards biocultural creatives. In: Arts BJM, Van Bommel S, Ros-Tonen MAF, Verschoor GM (eds) Forest-people interfaces; understanding community forestry and biocultural diversity. Wageningen Academic Publishers, Wageningen

Elands BHM, Wiersum KF, Buijs AE, Vierikko K (2015) Policy interpretations and manifestation of biocultural diversity in urbanized Europe: conservation of lived biodiversity. Biodivers Conserv 24:3347–3366

EPA (2018) Technical assistance document for the reporting of daily air quality – the air quality index (AQI). United States Environmental Protection Agency

Forman RTT (1995) Land mosaics: the ecology of landscape and regions. Cambridge University Press, Cambridge, UK

Forman RTT (2014) Urban ecology: science of cities. Cambridge University Press, Cambridge, UK

Fuller RA, Irvine KN, Devine-Wright P, Warren PH, Gaston KJ (2007) Psychological benefits of greenspace increase with biodiversity. Biol Lett 3(4):390–394

James P, Banay RF, Hart JE, Laden F (2015) A review of the health benefits of greenness. Curr Epidemiol Rep 2:131–142

Kaplan DH, Holloway S (2014) Urban geography, 3rd edn. Wiley, New York

Kim JE (2012) Green network analysis in coastal cities using least-cost path analysis: a study of Jakarta, Indonesia. J Ecol Environ 35:141–147

Kim JE (2019) Traditional ecological knowledge and sustainability of ecosystem services on islands: a case study of Shinan County, Jeollanamdo, Republic of Korea. J Marine Island Cult 8(1):28–35

Kim JE, Hong SK (2016) Pattern and process in MAEUL, a traditional Korean rural landscape. J Ecol Environ 34(2):237–249

Knobel P, Dadvand P, Alonso L, Costa L, Español M, Maneja R (2021) Development of the urban green space quality assessment tool (RECITAL). Urban For Urban Green 57:126895

Kusuma WL, Chih-Da W, Yu-Ting Z, Hapsari HH, Muhamad JL (2019) PM2.5 pollutant in Asia – a comparison of metropolis cities in Indonesia and Taiwan. Int J Environ Res Public Health 16: 4924

Kusumaningtyas SDA, Aldrian E, Wati T, Atmoko D, Sunaryo (2018) The resent state of ambient air quality in Jakarta, Aerosol Air Qual Res 18(9): 2343–2354

Lian Pin Koh, and Navjot S. Sodhi (2004) Importance of reserves, fragments, and parks for butterfly conservation in a tropical urban landscape. Ecol Appl 14(6):1695–1708. https://doi.org/https://doi.org/10.1890/03-5269

Maas J, Verheij RA, Groenewegen PP, De Vries S, Spreeuwenberg P (2006) Green space, urbanity, and health: how strong is the relation? J Epidemiol Community Health 60(7):587–592

Mabuhay J, Isagi Y (2005) Biodiversity in urban green space in Jabotabeck area, Indonesia. 2005 WSEAS Int. conf. on environment, ecosystems and development, Venice, Italy, November 2-4, 54–59

Maffi L (2005) Linguistic, cultural, and biological diversity. Annu Rev Anthropol 34:599–617

Maffi L, Woodley E (2010) Biocultural diversity conservation: a global sourcebook. Earthscan, London

Maheng D, Pathirana A, Zevengergen C (2021) A preliminary study on the impact of landscape pattern changes due to urbanization: case study of Jakarta. Indonesia Land 10:218. https://doi.org/10.3390/land10020218

Manan RH (2015) Effective management of green spaces Jakarta City. Int J Eng Res Technol 9(11):477–484

Manan RH (2016) Policy analysis of urban green open space management in Jakarta City, Indonesia. Int J Eng Res Technol 5(4):241–248

McGarigal K, Marks B (1995) FRAGSTATS: spatial pattern analysis program for quantifying landscape structure. Gen tech rep PNW-GTR-351, Portland, OR: US Department of Agriculture, Forest Service, Pacific northwest Research Station 122p

Nutsford D, Pearson AL, Kingham S (2013) An ecological study investigation the association between access to urban green space and mental health. Public Health 127(11):1005–1011

Ottitsch A, Krott M (2005) Urban Forest policy and planning. In: Konijnendijk CC, Nilsson K, Randrup TB, Schipperijin J (eds) Urban forests and tress: a reference book. Springer, Berlin, pp 117–148

Ramdhoni S, Rushayati SB, Prasetyo LB (2016) Open green space development priority based on distribution of air temperature change in capital city of Indonesia, Jakarta. Procedia Environ Sci 33:204–213

Rukmana D, Fahmi FZ, Firman T (2019) Suburbanization in Aisa: a focus on Jakarta. In: Hanlon B, Vicino TJ (eds) The Routledge companion to the suburbs. Routledge, London, UK, pp 110–120

Setiowati R, Hasibuan HS, Koesoer RH (2018) Green open space masterplan at Jakarta capital city, Indonesia for climate change mitigation. OP Conf Ser: Earth Environ Sci 200:012042

Shuvo FK, Feng X, Akaraci S, Astell-Burt T (2020) Urban green space and health in low and middle-income countries: a critical review. Urban For Urban Green 52:126662

Stålhammar S, Brink E (2020) 'Urban biocultural diversity' as a framework for human–nature interactions: reflections from a Brazilian favela. Urban Ecosyst. https://doi.org/10.1007/s11252-020-01058-3

Turner MG, Gardner RH, O'Neill RV (2001) Landscape ecology in theory and practice, Pattern and process. Springer, New York, p 401

Tzoulas K, Korpela K, Venn S, Wli-Pelkonen V (2007) Promoting ecosystem and human health in urban areas using green infrastructure: a literature review. Landsc Urban Plan 81(3):167–178

UNESCO (2019) The operational guidelines for the implementation of the World Heritage Convention, https://whc.unesco.org/en/guidelines/

United Nations (2021) World Population Prospects, https://population.un.org/wpp/

Van Bueren E, Van Bohemen H, Itard L, Visscher H (2014) Sustainable urban environments: an ecosystem approach. Springer, Dordrecht, The Netherlands

Van den Berg M, Wendel-Vos W, Van Poppel M, Kemper H, Van Mechelen W, Maas J (2015) Health benefits of green spaces in the living environment: a systematic review of epidemiological studies. Urban For Urban Green 14(4):806–816

Vierikko K, Elands B, Niemeiä J, Andersson E, Buijs A, Fischer LK, Haase D, Kabisch N, Kowarik I, Luz AC, Stahl AO, Száraz L, Van der Jagt A, Van der Bosch CK (2016) Considering the ways biocultural diversity helps enforce the urban green infrastructure in times of urban transformation. Curr Opin Environ Sustain 22:7–12

Vries D, Van Dillen SME, Groenewegen PP, Spreeuwenberg P (2013) Streetscape greenery and health: stress, social cohesion and physical activity as mediators. Soc Sci Med 94:26–33

WHO-Europe (2017) Urban green space interventions and health – a review of impacts and effectiveness. World Health Organization Regional Office for Europe, Copenhagen, Denmark

Zhang Y, Van den Berg AE, Van Dijk T, Weitkamp G (2017) Quality over quantity: contribution of urban green space to neighborhood satisfaction. Int J Environ Res Public Health 14(5):535

A Review on the Direction of Future Studies on Biocultural Landscapes in Forest and Agroforestry Systems in Indonesia

14

Amin Setyo Leksono and Aniza Zairina

Abstract

Indonesia is one of the mega-diverse countries with a very high diversity of species, ecosystems, and cultures. Among the diverse ecosystems on land, forests, agroforestry, and agricultural ecosystems provide many services to people's welfare. However, rapid changes in land use and the occurrence of disasters are among the major threats to those ecosystems. One of the disasters is landslide and it may create a critical bare land. This chapter presents a review of previous studies in Indonesia, including projections of future management and studies with a focus on good practices for biocultural landscape conservation and utilization of ecosystem services for human well-being. Moreover, it is divided into three parts such as covering the role of landscape agroforestry, the use of local biological resources, and conservation of critical lands. The chapter concluded that besides the potential and lots of services provided by forests and agroforestry systems, these ecosystems are prone to land-use change and natural hazards, which could negatively impact their performance. Therefore, efforts to overcome these negative impacts could be carried out with potential solutions, such as ecosystem reclamation, land revegetation, forest and agroforestry protection, and empowerment of community forests. This requires a more integrated approach in the interdisciplinary field of study, including the development of

A. S. Leksono (✉)
Department of Biology, Faculty of Mathematics and Natural Sciences, Universitas Brawijaya, Malang, East Java, Indonesia
e-mail: amin28@ub.ac.id

A. Zairina
Department of Biology, Faculty of Mathematics and Natural Sciences, Universitas Brawijaya, Malang, East Java, Indonesia

S. A. Abdullah et al. (eds.), *Conserving Biocultural Landscapes in Malaysia and Indonesia for Sustainable Development*,
https://doi.org/10.1007/978-981-16-7243-9_14

hydro-seeding methods for revegetation, evaluation of government policy implementation, and community participation.

Keywords

Agroforestry · Ecosystem service · Hydro-seeding · Local wisdom · Medicinal plant · Natural hazard · Revegetation

14.1 Introduction

Indonesia is one of the mega-diverse countries with a very high diversity of species, ecosystems, and cultures. Forest, agroforestry, and agricultural or plantation ecosystems are among its diverse terrestrial ecosystems and provide enormous services for the people's welfare. The concept of ecosystem services was divided into four categories from the time of its introduction by Daily in 1997 until the adoption by the United Nation Millennium Ecosystem Assessment in 2005. The categories are as follows, first, provision, which is in the form of products obtained from ecosystems such as food, fiber, and water. Second, regulation, which is in the form of benefits also derived from ecosystem processes, such as water and air filtration. Third culture, which is in the form of non-material benefits such as spiritual enrichment, cognitive development, recreation, and esthetic experiences, and forth, support which is in the form of ecological functions such as pollination, nutrient cycling, and soil formation.

The existence of various ecosystems with the unique local community culture forms a harmonious relationship between humans and nature (Leksono 2011). However, rapid land-use depletion, natural resource shortage, serious environmental pollution, ecosystem degradation, and natural disasters have led to biodiversity losses and reduction in the provision of ecosystem services. The rapid conversion of land uses particularly forest to cultivated land and settlements impedes the flow and services of ecosystems such as hydrological cycles, carbon sequestration, and others. This is exacerbated by frequent natural hazards such as floods and landslides which ultimately create critical bare lands (Marfai et al. 2008).

Most land-use conversion originated from anthropogenic activities. Moreover, the rapid population growth has led to an increase in various necessities of life including settlements and other infrastructures. The unbalanced ratio of the population with the extent of agricultural land has limited employment opportunities and income. This situation forces some people to commit violations of the forest, using marginal lands for their agricultural cultivation, and ignore the principle of environmental conservation (Wulandari 2010). Additionally, forest exploitation permits also play a major role in the conversion of forest landscapes into plantations, especially oil palm plantations. This is due to an imbalance between global demand, Indonesian timber, and palm oil production, which led to illegal or unsustainable logging and expansion of permanent agricultural areas between the mid-1990s and 2015 (Tsujino et al. 2016). In general, large-scale oil palm and timber plantations together account for more than two-fifths of national deforestation between 2001 and 2016. Moreover, oil palm plantations were the largest single driver of deforestation over that period, contributing 23% of the national deforestation (Austin et al. 2019). This corresponds closely to previous estimates in Indonesia (Austin et al. 2019). Similarly, Adrianto et al. (2020) reported that, from 1990 to 2017, secondary peat swamp forest decreased from 40,000 to 10,000 $km^{2,}$ and plantations including oil palm increased from about 10,000 to 40,000 km^2. This is because the dominant land-use transition was secondary peat swamp forest, which is directly converted into plantations or initially to shrubs and then into plantations.

Due to the massive changes in the landscape, the remaining forest, agroforestry, and agricultural or plantation landscapes need to be managed with a more sustainable approach. Therefore, conservation-based management is needed. The manifestation of this form of management needs to adopt three conservation functions, namely protection, preservation, and sustainable use. For this reason, this study addresses the good practices in the conservation of biocultural landscapes and the use of ecosystem services for human well-being.

Reviews of ecosystem services have been conducted at the service category level, for example, the cultural services category (Horcea-Milcu et al. 2013) and the ecological and economic service categories (Häyhä and Franzese 2014). The most recent review which was conducted by Xie et al. (2020) was presented in a more general form at the spatial level. In that paper, it was mentioned that the study of land ecosystem services in developed countries was led by the USA, while in developing countries it was led by China, Argentina, and Brazil. Furthermore, among the Southeast Asia countries, Thailand was solely mentioned as the number of articles published therein from 2000 to 2019 was nine. This indicates that studies on ecosystem services in Indonesia indexed in the Web of Science are still few. Lots of the studies are limited to local-level publications. Therefore, this chapter presents a review of previous studies on forest and, agroforestry services, values, and threats. It also suggests recommendations for future studies with a focus on good practices for the conservation of biocultural landscapes and the development of technology for land revegetation. This chapter is divided into three parts, covering the role of landscape agroforestry, utilization of local biological resources, and conservation of critical lands.

14.2 The Role and Value of the Agroforestry Ecosystem

Agroforestry in Indonesia is the dominant ecosystem particularly in mountainous areas in Java and Sumatra. It plays an important role in supporting land conservation and meeting the socio-economic needs of marginalized communities. Most of the agroforestry status is owned by the government and some are owned by the community. An agroforestry ecosystem is a form of multi-canopy land use consisting of a mixture of trees, shrubs, or seasonal plants which are often accompanied by livestock in one area of land. These ecosystems are generally bordered by forests. The livelihoods of rural communities around the forest are generally very dependent on natural resources, including forest products, both timber and non-timber (Birgantoro and Nurrochmat 2007). Based on government policy, forest and agroforestry management must be able to optimize the use of both ecosystems for the development of plantation and community forest (van Bodegom et al. 2008; De Royer et al. 2018). The approach aims to increase the added value and benefits of non-timber forest products and increase participation of the wider community in the development of plantation forests (Belcher 2005). In addition, the strategic forestry development plan is oriented towards the equitable balance of forestry development and economic growth of forest and land management communities. This can be achieved through the development of industrial and agribusiness crops in the forestry sector (Forestry Office of East Java Province 2010).

The existence and benefits of agroforestry have been widely studied, especially in Java and Sumatra (Safe'i 2014; Seeberg-Elverfeldt et al. 2009; van Noordwijk et al. 2016). Those studies can be classified into three groups including ecological, economic, and social studies. Ecological studies focus on various ecosystem services and vegetation profiles, social studies evaluate community perceptions, while economic studies evaluate the economic benefits of agroforestry services.

Ecological studies of agroforestry constitute various ecosystem services. Agroforestry services include controlling microclimate (Lin 2007; Hadi et al. 2016), preserving water (van Noordwijk et al. 2016), reducing landslides (Hairiah et al. 2006), reducing surface runoff and erosion, reducing nutrient losses through washing (Widianto and Lestari 2007; Suprayogo et al. 2002), community place attachment (Leksono et al. 2017), as well as maintaining the biodiversity of soil flora and fauna (Dewi 2007).

Agroforestry is composed of various types of trees and undergrowth intercrops that vary in planting system and age (Safe'i 2014). This system is relatively safer from the risk of crop failure and is more stable to market fluctuations and the effects of climate change (Gomes et al. 2020; Butarbutar 2012). There are various types of intercrops in the form of agroforestry, for example, konjack, cocoa, corn, and vegetables (Leksono et al. 2011). The intercrops managed by farmers are generally the type of cultivation that suits the type of soil and local climate and have important economic values. So, the ability to act as a buffer against biophysical, economic, and social changes is the main requirement for the conservation and sustainable use of agroforestry.

Agroforestry systems provide important economic and ecological benefits for farmers, one of which can provide income for farmers. From an economic aspect, agroforestry research focuses on economic benefits including creating diversified sources of income from the same land management (land sharing), through the agroforestry provision such as grass, medicinal plant, and intercrops (Olivi and Qurniati 2015; Kholifah et al. 2017) and potential of carbon sequestration services (Seeberg-Elverfeldt et al. 2009). A research conducted in Lampung showed the remarkable contribution of landscape agroforestry to community welfare. From that study, the contribution of agroforestry to income can reach more than 80 percent or IDR 50 to 100 million (about USD3500 to USD6900)/household/year (Olivi and

Qurniati 2015; Kholifah et al. 2017). The significant variables affecting farmer income include age, plantation area, number of workers, ethnicity, religion, the slope of the land, and credit assistance (Olivi and Qurniati 2015). Other studies have shown that farmers' income from plantation, crops, and husbandry in agroforestry system is two times greater (Kholifah et al. 2017). This shows that sustainable agroforestry management can support the economic welfare of the community.

From a social aspect, agroforestry research focuses on people's perceptions and knowledge to show their level of understanding and awareness towards agroforestry sustainable use. A study in Lampung showed that 42.07% of respondents have a good perception, while moderate perception was 28.28% and 29.65% respondents have a low level of perception. The significant factors for the people knowledge were education, the training frequency, and income (Wulandari 2010). Farmers have a high perception of agroforestry-patterned community forest management and the benefits of agroforestry-patterned community forests. The factors that correlate with farmers' perceptions are formal education, non-formal education, and the number of family dependents (Hudiyani et al. 2017; Kholifah et al. 2017).

In perspective of conservation, ecological and economic services can drive community awareness improvement on the benefits of agroforestry and promote better conservation practices. If farmers' income and environmental use are profitable, efforts to conserve ecosystems will increase, while land use changes will decrease.

Studies about agroforestry carried out in other countries also show the similar trend. Generally, those studies evaluate the benefits of agroforestry as carbon sequestration, buffering from flood and drought, biodiversity conservation, yield of goods and services to society, enhancing the fertility of the soils, and providing social and economic well-being to people (Shepherd and Montagnini 2001: Takimoto et al. 2008; Pandey 2007; Kumar 2010; Murthy et al. 2013; Quandt 2018). A study in Southern China shows that two typical agroforestry intercropping systems consisted of the paulownia (*Paulownia elongta*) intercropping system in the north and the tea (*Camellia sinensis*) in the south have higher energy efficiency and also a better financial benefit to farmers (Jianbo 2006). A more modern approach, which integrates modern management of pasture, forest, and cropped areas, can be further integrated to increase multiple uses of the territory and economic development. An assessment of landscape diversity value in Italy showed that agroforestry has higher value than the areas with specific land uses (Pardini 2009).

14.3 Utilization of Local Biological Resources in Forest and Agroforestry Ecosystem

Forests and agroforestry are home to diverse biological resources such as medicinal and edible plant species. Therefore, in Indonesia, most of the traditional medicines are harvested from wild plants (Hidayat 2012). Additionally, efforts to use medicinal plants are carried out by cultivating in agroforestry ecosystems (Hadi et al. 2016;

Iftitah and Haryono 2018). Most studies aim to invent a number of species and use them through traditional wisdom approaches.

Studies on the use of biological resources as medicinal plants have been extensively carried out in several places in Java, Sulawesi, Sumatra, and North Halmahera. A study in Kampung Naga village, West Java, reported 108 species of wild plants that were traditionally used for medicinal purposes by the community. The plants were obtained from the yards, forests, and agroforestry systems (Nurmalasari and Sukarsa 2012). The study in Gorontalo Province, Sulawesi showed 30 plant species used by the local community. They consisted of 24 medicinal plant species, two non-timber species, and four species of germplasm for other uses (Nurrani et al. 2015). Studies in Sumatra reported more than 194 plant species used by the local community. Meanwhile, a study in Halmahera reported 44 species (Yakub et al. 2019). Those plants were used to cure diseases such as wound, flatulence, constipation, seizures, high fever, puerperium, jaundice, diarrhea, high blood pressure, dry cough, vomiting, bitter yellow eyes, stiff and sore stomach, dizziness, sprains, asthma, and chills (Yani 2013).

Besides studies on medicinal plants, traditional wisdom studies were also carried out on edible plants for other purposes such as fuel, animal feed, wood, fruit, and others (Sher et al. 2011a, b; Pegu et al. 2013; Murad et al. 2013; Novita et al. 2021). These efforts aim to preserve local wisdom, management, utilization, promotion of sustainable livelihoods, food security, and ecosystem services (Yirga 2010; Somsri et al. 2017; Emiru et al. 2011).

Various studies on the species of fruit plants in yards and agroforestry systems showed that Indonesia has great potentials. A study in Aceh Tamiang District reported 39 species of fruit plants (Elfrida and Suwardi 2020), while the study in South Aceh District showed 56 species (Suwardi et al. 2020). Furthermore, the study in Karo District reported 30 species (Silalahi and Nisyawati. 2018). Native fruit plants play an important role in the daily life of people living in developing countries as they could be used for ornament purposes in cultural events (Mabaya et al. 2014; Khruomo and Deb 2018; Sutraningsih et al. 2019; Purwanto et al. 2020).

The practice of traditional wisdom is believed to be an effort to sustain the existence of multi-use wild plants (Tura et al. 2017). Moreover, the practice may support the conservation of biological resources through sustainable use. The concept of sustainable use arises from the principle of direct harvesting of wild plants at a rate that surpasses the limits of their regeneration capacity in nature (Russell et al. 2015). This effort could therefore provide benefits both economically and ecologically. Economically, the use of biological resources could support income for local people, while the ecological benefits are generally associated with the efforts to preserve the habitat for medicinal and food plants. By preserving the habitat, other biological resources and the environment are also preserved. Another advantage of the efforts is the bioprospecting of germplasm for future breeding programs (Purwanto et al. 2020).

14.4 Conservation of Critical Ecosystems

Beside posses important ecosystem services, however, its sustainability is often threatened by land conversion. Forest and agroforestry are prone to potential natural disasters. The geographic location of these ecosystems is vulnerable to floods, landslides, and volcanic eruptions. These natural hazards may destruct the ecosystem and even create bare land. The threats may reduce its services, for example, the opening of agroforestry ecosystems into pure moors or settlements has an impact on reducing runoff water uptake, and carbon sequestration.

Post-disaster handling is generally carried out by constructing a retaining embankment or leaving it withstand landslide scars if the location is away from a road or building. The abandoning post landslide lands slows down the revegetation process because landslides have eroded the entire topsoil and left poor soil, resulting in the formation of critical land. Besides, critical land can also be caused by human activities, especially mining activities.

Critical land is degraded land that has lost its service function as water regulator, agricultural production supply, and environmental protection elements. Critical lands were occurred due to degradation process of physical, chemical, and biological soil characteristics (Lal 2015; Didu 2011). The restoration of critical land involves the activities and policies of various institutions that are interrelated. Although efforts to rehabilitate land have been carried out through reforestation and other programs, these have not been able to cope with the extent of critical land area yet (Didu 2011). Revegetation efforts are frequently unfeasible in the field due to accessibility, damaged land, community participation, weather, funding, etc. Critical land generally occurs in areas with low accessibility, steep, and fragile.

Various methods to address land accessibility continue to be developed, one of which is replanting using seedlings or developing technology, such as hydro-seeding. This chapter describes the hydro-seeding method and the follow-up of post-seedling plant care. This technique has been widely applied to large and steep slopes in temperate climates. According to Clemente, hydro-seeding is often used for the restoration of critical land, for example, ex-mining land (Clemente et al. 2016). Hydro-seeding is a technique that can be applied to land revegetation. Hydro-seeding technology is a planting process using a mixture or formula consisting of plant seeds, soil stabilizer, water, and mulch (Parsakhoo et al. 2018). The formula is transported in a tank or truck. Then the formula is sprayed over the prepared slope area in a uniform site. The hydro-seeding method is an alternative technology that is more effective and efficient than the manual method. Hydro-seeding with commercial seeds of fast-growing grasses and legumes is a common practice in revegetation. Several previous studies have always been associated with ex-mining land (Thomas et al. 2015; Yulianingsih and Arisoesilaningsih 2015). Meanwhile, hydro-seeding is one of the revegetation actions that make it possible to overcome the difficulties of manual planting. The hydro-seeding formula can be applied in a relatively wide and steep area (Albaladejo et al. 2000). The presence of an adhesive in the formula increases the ability of the seeds to resist erosion.

The choice of plant species is very important and must fulfill the purpose for which the species is designated to ensure that the slope stability as a target is in harmony with the surrounding environment. For example, the grass species are preferable because this plant profile may cover the soil quickly (Zairina and Mondiana 2020). Grass and legumes are the types commonly planted in the early stages of land revegetation because they are pioneer species (Chen et al. 2014). After several generations, the leaf litter produced by these plants can reduce the rate of erosion and the speed of ground runoff.

Several studies promote the use of indigenous plant and avoid alien species. The use of non-native plants may potentially dominate the community and some cases become invasive (García-Palacios et al. 2010). Therefore, native plant species are highly recommended because of their presumed ability to adapt to local abiotic and biotic environments. Several types of grass species and local legumes such as *Cyperus brevifolius*, *C. eragrostis*, *C. odoratus*, *C. strigosus*, *Kyllingia monocephala*, *Centrosema pubescens*, *Pueraria javanica*, and *Calopogonium mucunoides* are species that received a lot of attention and continue to be studied more intensively (Bargawa et al. 2019; Anshari et al. 2018; Yulianingsih and Arisoesilaningsih 2015; Zairina and Mondiana 2020). The stability of the slope as the target should be in harmony with the surrounding environment. This goal may be successful with revegetation, using native species (Stokes et al. 2009; Clemente et al. 2016).

Grass seeds for hydro-seeding formula have a very small size and mixed with complementary materials. These materials consist of a growing medium and adhesive material. Both function to provide nutrients and to bind the seeds onto the soil. With a strong bond it would protect the seed from erosion by rain-water. Previous research generally used commercial grasses and legumes materials. Commercial mixtures are usually successful in producing a plant cover over the first few years, but often fail in the medium-long term, since a self-sustaining vegetation cover is only guaranteed through the use of ecologically adapted species (Krautzer et al. 2013). The use of hydro-seeding formula with commercial latex and PAM soil stabilizer types in the mixture remains widely used because it is considered to have less effect on growth. The use of mulch and compost also greatly influenced the growth of pioneer plants. Organic compost is thought to have a positive impact because it is more environmentally friendly. Rice husk is a good companion material compared to rice straw and sawdust. With a mixture of mulch from buffalo fecal compost grown on mining soil media had the highest plant density compared to other media (Yulianingsih and Arisoesilaningsih 2015).

14.5 Framework of Future Management and Study

Based on the description of the review, this chapter proposed a framework for the scientific development of forest and agroforestry landscapes conservation strategies through studies and research. As a country with a high level of biodiversity, Indonesia has the potential for its natural resources to provide a very diverse range of ecosystem services. However, there are lots of threats from nature such as natural disasters or anthropogenic activities. These events could put pressure on the sustainability of ecosystem services.

To overcome the negative impacts, potential solutions could be taken including ecosystem reclamation, land revegetation, forest and agroforestry protection, and community forest empowerment. The role of research in the provision of potential solutions covers at least a variety of topics. This chapter focused on suggesting three topics, namely the development of hydro-seeding methods, policy implementation, and community participation.

First, one of the efforts in revegetation could be carried out using the hydro-seeding method. Given the increasing extent of critical land, research on the development of this method has great potential in the future. Finally, the study of native plant species growth, compost materials, and organic soil stabilizers also needs attention because of the abundance of local potentials.

Second, regarding the implementation of government policies that are oriented towards sustainable use, the opportunity to study the impact of policies on an economic, social, and ecological basis is available. This is also inseparable from the new direction of the Indonesian government's policy through ombudsman law which is more open to investment and encourages forest use by local communities. On the one hand, this policy would have a positive impact on economic growth, however, it could also increase negative environmental impacts.

Third, efforts to minimize the rate of land-use change and ecosystem destruction could be carried out through structural efforts such as reforestation, afforestation, and maintenance of dams and terraces, including non-structural efforts such as involving the community, increasing income, and others. Moreover, these efforts need to be carried out in an integrated and sustainable manner. Experience has proven that management efforts carried out separately by each sector, without the basis of an interdisciplinary or integrated approach, often lead to conflicts between sectors, for example, between the forestry sector and the agricultural sector. Meanwhile, by using an interdisciplinary or integrated approach in land use and regional planning, these conflicts could be avoided (Nugroho et al. 2019). Therefore, the study of the revegetation process using the hydro-seeding method needs to be integrated with community participation.

Hydro-seeding is a specific method for grass and legume group planting. After these plants have grown, further revegetation needs to be carried out by planting tree seedlings. Some revegetation efforts with tree species require community participation including in providing funds for maintenance. Therefore, some programs were introduced, and an example is the tree adoption program. This program was carried out through a donation process for tree maintenance. The results of the donation were then used for the maintenance or embroidery of the dead trees in more accessible areas. In this concern, this chapter proposed studies on the need for community participation in long-term revegetation to be explored in the future (Fig. 14.1).

Forest community empowerment through education and training has a significant impact on the restoration of degraded lands. Moreover, the existence of forest

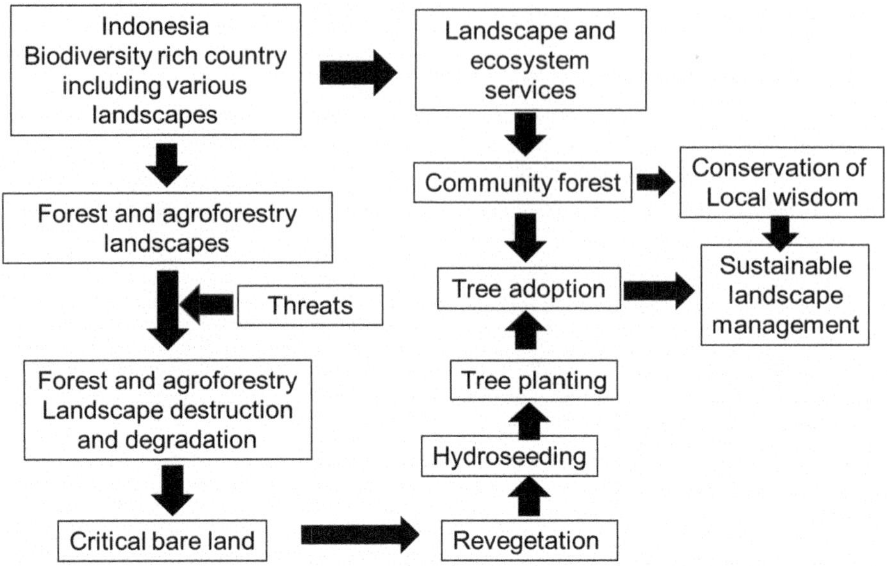

Fig. 14.1 Framework of future forest and agroforestry landscape management and studies

communities improves awareness of forest functions, which promotes effective practice for the success of land revegetation. Community development based on human resources, socio-culture, economy, community institutional structures, natural resources, environmental tourism, inter-institutional harmonization needs to be encouraged because these are the right strategies for conserving forest biodiversity (Nugroho et al. 2019). For this reason, good practices in managing biological resources need to be maintained and preserved through the actualization of local wisdom values and support from parties in developing technology that is more environmentally friendly. This is inseparable from community participation in jointly supporting efforts in the use of ecosystem services in a sustainable manner (Fig. 14.1).

The achievements in the preservation of biological resources and ecosystems are marked by the ability of the ecosystem to support human life and the environment. In other words, a sustainable ecosystem is an ecosystem that could carry out all its functions. Conservation efforts at the landscape scale mean that ecosystem preservation needs to be an inseparable part of human activities. Therefore, policy support from the government, community participation, and support from the private sector is needed to realize this conservation effort.

14.6 Conclusion

This chapter concludes that though forests and agroforestry systems have great potential and lots of ecosystem services, they are also prone to land-use change and natural hazards. These threats could have a negative impact on the performance of the ecosystems. Therefore, efforts to overcome the negative impacts such as ecosystem reclamation, land revegetation, forest and agroforestry protection, and community forest empowerment need to be carried out. This requires a more integrated approach in the interdisciplinary field of study.

The success of the above efforts could be measured from three healthy and sustainable ecosystem functions. First, ecological function, this means that the ecosystem is able to guarantee its sustainability and there is a balanced interaction between its components. Second, economic function, which means that the ecosystem is able to support the economic welfare of the community, and third, social functions, and this could be achieved by creating an empowered and independent society. Therefore, research in the future needs to consider these aspects. The role of research to discover potential solutions covers at least a variety of topics. Therefore, this chapter recommended three scopes of study, as follows: the development of hydro-seeding methods, policy implementation, and community participation.

References

Adrianto HA, Spracklen DV, Arnold SR, Sitanggang IS, Syaufina L (2020) Forest and land fires are mainly associated with deforestation in Riau Province, Indonesia. Remote Sens 12(1):3. https://doi.org/10.3390/rs12010003

Albaladejo J, Alvarez-Rogel J, Querejeta J, Diaz-Pereira E, Castillo V (2000) Three hydro-seeding revegetation techniques for soil erosion control on anthropic steep slopes. Land Degrad Dev 11(4):315–325

Anshari MF, Fernandes AAR, Arisoesilaningsih E (2018) Comparing seeds germination of some local plant species on two hydroseeding mulches for post-mining revegetation. J Degraded Mining Land Manag 5(2):1103–1110. https://doi.org/10.15243/jdmlm.2018.052.1103

Austin KA, Schwantes A, Gu Y, Kasibhatla PS (2019) What causes deforestation in Indonesia? Environ Res Lett 14(2):1103–1110

Bargawa WS, Putra A, Nurcholis M (2019) Analysis of erosion using hydroseeding on post-coal mining in Melak site. Intl J Geomate 17:371–377

Belcher BM (2005) Forest product markets, forests and poverty reduction. Int For Rev 7(2):82–89

Birgantoro BA, Nurrochmat DR (2007) Forest resource utilization by people in KPH Banyuwangi Utara. JMHT Vol. XII. Jurnal Manajemen Hutan Tropika 3:172–181

van Bodegom A, Berg J, van der Meer P (2008) Forest plantations for sustainable production in the tropics: key issues for decision-makers. Wageningen University & Research Centre, The Netherlands

Butarbutar T (2012) Agroforestri untuk adaptasi dan mitigasi perubahan iklim. Jurnal Analisis Kebijakan Kehutanan 9(1):1–10

Chen YC, Wu CF, Lin SH (2014) Mechanisms of Forest restoration in landslide treatment areas. Sustainability 6(10):6766–6780. https://doi.org/10.3390/su6106766

Clemente AS, Moedas AR, Oliveira G, Martins-Loução MA, Correia O (2016) Effect of hydroseeding components on the germination of Mediterranean native plant species. J Arid Env 125:68–72

De Royer S, Van Noordwijk M, Roshetko JM (2018) Does community-based Forest Management in Indonesia Devolve Social Justice or social costs? Int For Rev 20(2):167–180. https://doi.org/10.1505/146554818823767609

Dewi WS (2007) Dampak alih guna lahan hutan menjadi lahan Pertanian: Perubahan diversitas cacing tanah dan fungsinya dalam mempertahankan pori makro tanah. Paska Sarjana, Universitas Brawijaya, Malang, Disertasi

Didu MS (2011) Analisis posisi dan peran lembaga serta kebijakan dalam proses pembentukan lahan kritis. Jurnal Teknologi Lingkungan 2(1):93–105

Elfrida MA, Suwardi AB (2020) The fruit plant species diversity in the home gardens and their contribution to the livelihood of communities in rural area. Biodiversitas 21(8):3670–3675

Emiru B, Ermias A, Wolde M, Degitu E (2011) Management, use, and ecology of medicinal plants in the degraded drylands of Tigray. Northern Ethiopia J Hortic For 3(2):32–41

García-Palacios P, Soliveres S, Maestre FT, Escudero A, Castillo-Monroy AP, Valladares F (2010) Dominant plant species modulate responses to hydroseeding, irrigation and fertilization during the restoration of semiarid motorway slopes. Ecol Eng 36(10):1290–1298

Gomes LC, Bianchi FJJA, Cardoso IM, Fernandes RBA, Filho EIF, Schulte RPO (2020) Agroforestry systems can mitigate the impacts of climate change on coffee production: a spatially explicit assessment in Brazil. Agric Ecosyst Environ 294:1–11

Hadi EEW, Widyastuti SM, Wahyuono S (2016) Diversity and Untilization of Understorey in agroforestry system of Menoreh Hill, Kulon Progo regency. Jurnal Manusia dan Lingkungan 23(2):206–215

Hairiah K, Sulistyani H, Suprayogo D, Widianto, P. Purnomosidhi, Widodo RH, and Van Noordwijk M, (2006). Litter layer residence time in forest and coffee agroforestry systems in Sumberjaya, West Lampung. Forest Ecol Manag 224: 45–57

Häyhä T, Franzese PP (2014) Ecosystem services assessment: a review under an ecological-economic and systems perspective. Ecol Model 289:124–132

Hidayat S (2012) Existence of endangered medicinal plant and its uses in Bogor surrounding areas. Media Kesehatan 17(1):33–38

Hudiyani I, Purnaningsih N, Asngari PS, Hardjanto H (2017) Persepsi Petani terhadap Hutan Rakyat Pola Agroforestri di Kabupaten Wonogiri. Provinsi Jawa Tengah Jurnal Penyuluhan 13(1):64–78. https://doi.org/10.25015/penyuluhan.v13i1.14709

Iftitah SN, Haryono G (2018) Pengkajian Beberapa Tanaman Empon - Empon di Desa Balesari Kecamatan Windusari Kabupaten Magelang. Jurnal Ilmu Pertanian Tropika dan Subtropika 3(1):13–16

Jianbo L (2006) Energy balance and economic benefits of two agroforestry systems in northern and southern China. Agric Ecosyst Environ 116:255–262

Kholifah UN, Wulandari C, Santoso T, dan Kaskoyo H. (2017) Kontribusi agroforestri terhadap pendapatan petani di Kelurahan Sumber Agung Kecamatan Kemiling Kota Bandar Lampung. Jurnal Sylva Lestari 5(3):39–47

Khruomo N, Deb CR (2018) Indigenous wild edible fruits: sustainable resources for food, medicine and income generation-a study from Nagaland. India J Exp Biol Agric Sci 6(2):405–413

Krautzer B, Graiss W, Klug B (2013) Ecological restoration of ski runs. In: Rixen C, Rolando A (eds) The impacts of skiing and related winter recreational activities on mountain environments. Bentham Ebooks

Kumar AK (2010) Carbon sequestration: underexploued environmental benefits of Tarai agroforestry. Indian J Soil Con 38:125–131

Lal R (2015) Restoring soil quality to mitigate soil degradation. Sustainability 7:5875–5895. https://doi.org/10.3390/su7055875)

Leksono AS (2011) Keanekaragaman Hayati: teori dan aplikasi. UB Press

Leksono AS, Yanuwiadi Y, Zaenal K, Farid A, Maulana F (2011) Influence of Porang (Amorphophalus muelleri) cultivation on the composition of soil arthropods in tropical agroforestry areas in East Java. Indonesia J Trop Life Sci 1(2):76–81

Leksono AS, Poewoningsih D, Ika NWP (2017) Green open space demand and community place attachment in Batu, East Java. In: Hong SK, Nakagoshi N (eds) Landscape ecology for sustainable society. Springer, Cham

Lin B (2007) Agroforestry management as an adaptive strategy against potential microclimate extremes in coffee agriculture. Agricultural and Forest Meteorology - AGR FOREST METEOROL 144:85–94. https://doi.org/10.1016/j.agrformet.2006.12.009

Mabaya E, Jackson J, Ruethling G, Carter CM, Castle J (2014) Wild fruits of Africa: commercializing natural products to improve rural livelihoods in southern Africa. Intl Food Agribus Man 17B(B):69–74

Marfai MA, King L, Singh LP et al (2008) Natural hazards in Central Java Province, Indonesia: an overview. Environ Geol 56:335–351. https://doi.org/10.1007/s00254-007-1169-9

Milcu AI, Hanspach J, Abson D, Fischer J (2013) Cultural ecosystem services: a literature review and prospects for future research. Ecol Soc 18(3):44. https://doi.org/10.5751/ES-05790-180344

Murad W, Azizullah A, Adnan M, Akash T, Saqib W, Ashfaq A (2013) Ethnobotanical assessment of plant resources of Banda Daud Shah, district Karak. Pakistan J Ethnobiol Ethnomed 9:77. https://doi.org/10.1186/1746-4269-9-77

Murthy IK, Gupta M, Tomar S, Munsi M, Tiwari R et al (2013) Carbon sequestration potential of agroforestry Systems in India. J Earth Sci Climate Change 4:131. https://doi.org/10.4172/2157-7617.1000131

van Noordwijk M, Kim YS, Leimona B, Hairiah K, Fisher LA (2016) Metrics of water security, adaptive capacity, and agroforestry in Indonesia. Curr Opin Environ Sustain 21:1–8

Novita K, Indriyani S, Arumingtyas E, Azrianingsih R (2021) Local snake fruit conservation in East Java, Indonesia: community knowledge and appreciation Biodiversitas 22(1): 416–423

Nugroho M, Soemarno, Riniwati H, Afandhi A (2019) Forest vegetation profile for springs preservation in Arjuna Mount, East Java-Indonesia. Am J Agric For 7(6):290–296. https://doi.org/10.11648/j.ajaf.20190706.17

Nurmalasari N, Sukarsa HHA (2012) Studi Kasus Pemanfaatan Tumbuhan sebagai Obat-Obatan Tradisional oleh Masyarakat Adat Kampung Naga di Kabupaten Tasikmalaya. Biosfera 29(3):141–150

Nurrani L, Tabba S, Mokodompit HS (2015) Kearifan lokal dalam pemanfaatan tumbuhan obat oleh masyarakat di sekitar Taman Nasional Aketajawe Lolobata, Provinsi Maluku Utara. Jurnal Penelitian dan Sosial Ekonomi Kehutanan 12(3):163–175

Olivi R, Qurniati R, Firdasari (2015). Kontribusi Agroforestri Terhadap Pendapatan Petani Di Desa Sukoharjo 1 Kecamatan Sukoharjo Kabupaten Pringsewu. Jurnal Sylva Lestari 3(2): 1–12

Pandey DN (2007) Multifunctional agroforestry systems in India. Curr Sci 92:455–463

Pardini A (2009) Agroforestry Systems in Italy: traditions towards modern management. In: Rigueiro-Rodróguez A, McAdam J, Mosquera-Losada MR (eds) agroforestry in Europe. Advances in agroforestry, vol 6. Springer, Dordrecht. doi:https://doi.org/10.1007/978-1-4020-8272-6_12

Parsakhoo A, Jajouzadeh M, Motlagh RA (2018) Effect of hydroseeding on grass yield and water use efficiency on forest road artificial soil slopes. J For Sci 64:157–163

Pegu R, Gogoi J, Tamuli AK, Teron R (2013) Ethnobotanical study of wild edible plants in Poba reserved Forest, Assam, India: multiple functions and implications for conservation. Res J Agric Fores Sci 1(3):1–10

Purwanto SR, Novenda IL, Narulita E (2020) Ethnobiology study on nature and processing plants and animals in Traditional Sea-picking of Banyuwangi regency. Mangifera Edu 5:38–54. https://doi.org/10.31943/mangiferaedu.v5i1.90

Quandt A (2018) Measuring livelihood resilience: the household livelihood resilience approach (HLRA). World Dev 107:253–263

Russell J, Yates CP, Edwards AC, Whitehead PJ, Murphy BP, Lawes MJ (2015) Deriving multiple benefits from carbon market-based savanna fire management: an Australian example. PLoS One 10(12):e0143426. https://doi.org/10.1371/journal.pone.0143426

Safe'i R (2014) Value of vitality status in monoculture and agroforestry planting Systems of the Community Forests. Int J Sci Basic Appl Res (IJSBAR) 18(2):340–353

Seeberg-Elverfeldt C, Schwarze S, Zeller M (2009) Payments for environmental services – carbon finance options for smallholders' agroforestry in Indonesia. Int J Commons 3(1):108–130

Shepherd D, Montagnini F (2001) Above-ground carbon sequestration potential in mixed and pure tree plantations in the humid tropics. J Trop For Sci 13:450–459

Sher H, AlYamani MN, Wijaya L (2011a) Ethnobotanical and antibacterial potential of Salvadora persica: a well-known medicinal plant in Arab and union system of medicine. J Med Plants Res 5(7):1224–1229

Sher Z, Khan ZUD, Hussain F (2011b) Ethnobotanical studies of some plants of Chagaharzai valley, district Buner. Pakistan Pak J Bot 43(3):1445–1452

Silalahi S, Nisyawati. (2018) The ethnobotanical study of edible and medicinal plants in the home garden of Batak Karo sub-ethnic in North Sumatra. Indonesia. Biodiversitas 19(1):229–238

Somsri S, Yoovatana M, Noppornphan C, Sangudom T, Palakorn K, Prasongsup S, Sripinta P, Posawang S, Nimkingrat T, Rukkhaphan A, Siriyan R, Changprasert S, Tongtao S, Dum-ampai N, Ariyapuchai S, Issarakraisila M, Changtragoon S, Rungrattanakul P (2017) Conservation and sustainable use of cultivated and wild tropical fruit diversity: promoting sustainable livelihoods, food security and ecosystem services. Acta Hortic 1186:69–76. https://doi.org/10.17660/ActaHortic.2017.1186.11

Stokes A, Atger C, Bengough AG, Fourcaud T, Roy CS (2009) Desirable plant root traits for protecting natural and engineered slopes against landslides. Plant Soil 324:1–30. https://doi.org/10.1007/s11104-009-0159-y

Suprayogo D, Van Noordwijk M, Hairiah K, Cadisch G (2002) The inherent 'safety-net' of ultisols: measuring and modeling retarded leaching of mineral nitrogen. Eur J Soil Sci 53:185–194

Sutraningsih NKA, Sukenti K, Sukiman AE (2019) Ethnobotanical study on Daksina constituent plants on Lombok Island, West Nusa Tenggara. Indonesia Asian J Ethnobiol 2(2):48–53

Suwardi AB, Navia ZI, Harmawan T, Syamsuardi ME (2020) Ethnobotany and conservation of indigenous edible fruit plants in South Aceh. Indonesia Biodiversitas 21(5):1850–1186

Takimoto A, Nair PKR, Nair VD (2008) Carbon stock and sequestration potential of traditional and improved agroforestry systems in the west African Sahel. Agric Ecosyst Environ 125:159–166

Thomas C, Sexstone A, Skousen J (2015) Soil biochemical properties in brown and gray mine soils with and without hydroseeding. Soil 1:621–629. https://doi.org/10.5194/soil-1-621-2015

Tsujino R, Yumoto T, Kitamura S, Djamaluddin I, Darnaedi D (2016) History of forest loss and degradation in Indonesia. Land Use Policy 57:335–347

Tura TT, Soromess T, Argaw M, Leta S (2017) Holy hills role in biodiversity conservation and mountain management: implication of multi-purpose Ethiopian orthodox Tewahido church in sustainable mountain conservation. J Biodivers Environ Sci 1:179. https://doi.org/10.4172/2332-2543.1000179

Widianto SD, Lestari ID (2007) Alih Guna Lahan Hutan Menjadi Lahan Pertanian: Apakah fungsi hidrologis hutan dapat digantikan sistem kopi monokultur ? Prosiding Seminar sehari: "Penanganan Bencana Sumber Daya Pertanian", 1 Februari 2007. Universitas Brawijaya, Malang, Fakultas Pertanian

Wulandari C (2010) Studi persepsi masyarakat tentang pengelolaan lanskap agroforestry di sekitar Sub DAS Way Besai Provinsi Lampung. Jurnal Ilmu Pertanian. Indonesia:15137–15140

Xie H, Zhang Y, Choi Y, Li F (2020) A Scientometrics review on land ecosystem service research. Sustainability 12. https://doi.org/10.3390/su12072959

Yakub A, Leksono AS, Batoro J (2019) Ethnobotany of medicinal and edible plants of Tobelo Dalam tribe in Aketajawe Lolobata National Park Area. Jurnal Pembangunan dan Alam Lestari 10(1):45–50

Yani AP (2013) Kearifan Lokal Penggunaan Tumbuhan Obat Oleh Suku Lembak Delapan Di Kabupaten Bengkulu Tengah. Bengkulu Prosiding Semirata FMIPA Universitas Lampung:71–74

Yirga G (2010) Assessment of indigenous knowledge of medicinal plants in central zone of Tigray, northern Ethiopia. African J Plant Sci 4:6–11

Yulianingsih D, Arisoesilaningsih E (2015) Aplikasi Beberapa Mulsa Hydroseeding untuk Perkecambahan Biji Teki Pioner di Tanah Pasca Pertambangan Batubara dari Kalimantan Selatan. J Biotrop 3(1):27–31

Zairina A, Mondiana YQ (2020) The effect of slope level and hydroseeding formula on the growth of *Cyperus brevifolius* and *Pueraria javanica* seedlings. Biodiversitas 21(11):5008–5012

Conserving Biocultural Landscapes: The Need for Sustainable Development

15

Saiful Arif Abdullah and Amin Setyo Leksono

Abstract

The biocultural landscape is a heritage demonstrating the long-standing close relationship between nature and culture. This heritage is valuable and meaningful to a property or area because it symbolises harmony between humans and between humans and the environment. In this regard, harmony demonstrates the concept of sustainability, which promotes the well-being of the environment, economy and society; this is the core characteristic of a biocultural landscape. Thus, biocultural landscape conservation is important and necessary for the sustainable development of a country. This chapter outlines the importance and need for conserving the biocultural landscape in Malaysia and Indonesia, supporting sustainable development in both countries. This chapter explains the recognition of natural and cultural heritage to enable its holistic conservation within the global sustainable development agenda and to support achieving the Sustainable Development Goals by 2030. This is important for understanding the need to conserve the biocultural landscape at the national or regional level, which is relevant to the global agenda of sustainable development.

S. A. Abdullah (✉)
Institute for Environment and Development (LESTARI), Universiti Kebangsaan Malaysia, Bangi, Selangor Darul Ehsan, Malaysia
e-mail: saiful@ukm.edu.my

A. S. Leksono
Department of Biology, Faculty of Mathematics and Natural Sciences, Universitas Brawijaya, Malang, East Java, Indonesia
e-mail: amin28@ub.ac.id

S. A. Abdullah et al. (eds.), *Conserving Biocultural Landscapes in Malaysia and Indonesia for Sustainable Development*,
https://doi.org/10.1007/978-981-16-7243-9_15

Keywords

Biocultural landscape · Heritage conservation · Malaysia · Indonesia · Sustainability · Sustainable Development Goals · Traditional heritage

15.1 Introduction

Malaysia and Indonesia are rich in biological and cultural diversity. In terms of biological diversity or biodiversity, both countries are amongst the 17 most mega-diverse countries in the world. In addition, both countries have various ethnic or tribal groups that influence and shape their cultural diversity. Over a long time, these biological and cultural components have built close or interdependent relationships with each other. These relationships are demonstrated primarily via the sustainable use of biological resources through traditional (i.e. cultural) practices that emphasise social, economic and environmental benefits. Biological resources are used as a source of food, as well as medical practices, traditional ceremonies or events, fine arts, architecture and agriculture. This creates a unique, complex socio-ecological system specific to a certain location known as a biocultural landscape.

The uniqueness of a biocultural landscape is a heritage that requires commitment from various stakeholders, including local communities, researchers (science and social science), conservationists, planners and decision makers, to ensure its current and future sustainability. In this context, conserving a biocultural landscape is important for supporting the sustainable development of both Malaysia and Indonesia. Therefore, to understand the importance of biocultural landscape conservation within the sustainable development agenda, this chapter first describes how natural and cultural heritage is recognised to enable its holistic conservation within the global sustainable development agenda and achieve the Sustainable Development Goals by 2030. This discussion demonstrates the need to conserve biocultural landscapes at the national or regional level relevant to global sustainable development efforts. Then, we outline the need and importance of biocultural landscape conservation for sustainable development in Malaysia and Indonesia and provide overall conclusions.

15.2 Natural and Cultural Heritage in the Global Sustainable Development Agenda

Historically, the commitment to conserving biological (hereafter referred to as natural) and cultural heritage began in the early 1970s. This was underlined in a convention related to protecting world natural and cultural heritage, which is better known as the World Heritage Convention by the United Nations Educational, Scientific and Cultural Organization (UNESCO; UNESCO 1972). The UNESCO General Conference adopted the convention in 1972, but it did not come into force until 1975 when it became ratified by 20 nations (Slatyer 1983). The signing of the convention was mainly driven by international community concern particularly heritage conservationists, about the deterioration of natural and cultural heritage as a result of rapid socio-economic growth (Labadi 2014). To date, the convention has been ratified by 194 States Parties, including Malaysia and Indonesia. The convention is considered a global flagship for the conservation, preservation and management of natural and cultural heritage.

For each rectified country, the World Heritage Convention is an important instrument for inventorying, identifying, protecting and preserving areas or places with a distinctive feature of global heritage (UNESCO 1972). The convention has two distinct domains: natural and cultural heritage. Natural heritage includes natural features, geological and physiographical formations and natural sites, whereas cultural heritage includes monuments, buildings and sites (UNESCO 1972). Generally, the convention's approach to conserving natural heritage is distinct from that for cultural heritage, as indicated within the UNESCO World Heritage framework. There is no integration or interrelationship between the two domains in the conservation framework; however, the approach for both domains maintains the spirit of sustainable development, with the main purpose of preserving heritage. Serious discourse on sustainable development only began in the early 1980s. It officially or institutionally became a global environmental agenda in 1987 upon the publication of the Brundtland Report by the World Commission on Environment and Development (Elliot 2006). In line with the emergence of sustainable development as being at the forefront of the global development agenda, since the 1980s, many parties have realised that the natural-cultural dichotomous approach is not conducive to effective heritage conservation (Chapagain 2019).

In recent times, the natural and cultural heritage conservation discourse has emphasised a holistic approach to address the gaps inherent to the natural-cultural dichotomous approach. Thus, the concept of cultural landscape received more attention, even though the idea began much earlier, around the second half of the twentieth century. This is portrayed through the introduction of the cultural landscape as a new category in the Operational Guidelines 1994, and the notion of 'sustainability' was used to refer to the sustainable use of cultural landscape (UNESCO 1994). In short, the concept of cultural landscape recognises human interaction with the landscape and vice versa, in which landscape shapes human culture and contributes to a local identity. Even though the term 'biocultural landscape' is not explicitly depicted, cultural landscape in a wider context (see Chap. 1 and Chapagain 2019) encompasses biotic or natural resources; it is not limited to cultural elements. In fact, a cultural landscape itself includes biotic resources such as wildlife and forests, the elements that also characterise or shape a natural landscape. Therefore, natural and cultural heritage conservation has evolved to be more inclusive, encompassing a wider scope and context without separating the two domains.

In addition, a series of sessions by the World Heritage Committee has also emphasised sustainability and sustainable development in the context of the conservation and management of natural and cultural heritage. One example is the Budapest Declaration, signed in 2002, which emphasised the need for a balance between conservation, sustainability and development that ultimately contributes to socio-economic development and the well-being of society (UNESCO 2002). Meanwhile, during the 31st session, the committee added the role of community into the implementation of the convention (UNESCO 2007); this addition recognised the importance of community involvement in the conservation of natural and cultural heritage, which is one of the features of sustainable development. During the 35th session, the committee further strengthened the application of sustainable development principles in its Operational Guidelines, including the full involvement of stakeholders in the identification, protection and management of heritage properties (UNESCO 2011a). Additionally, in 2011, the Strategic Action Plan for the Implementation of the Convention, 2012–2022 was adopted, which also integrated the idea of sustainable development (UNESCO 2011b). Recently, the *Policy on the Integration of a Sustainable Development Perspective into the Processes of the World Heritage Convention was adopted in 2015,* during the 20th General Assembly of the States Parties to the World Heritage Convention (UNESCO 2015).

Efforts to conserve natural and cultural heritage do not lie solely in the hands of the UNESCO World Heritage Convention. Other international institutions, including the International Centre for the Study of the Preservation of Cultural Property (ICCROM), the International Council on Monuments and Sites (ICOMOS) and the International Union for Conservation of Nature (IUCN), provide support. These institutions cooperate with the UNESCO World Heritage Convention and have their own programmes, projects or activities in support of the direction. For example, ICOMOS recently adopted the Heritage and the Sustainable Development Goals: Policy Guidance for Heritage and Development Actors, which provides direction regarding how heritage can handle the Sustainable Development Goals (SDGs, Labadi et al. 2021). Interestingly, one of the calls under the policy direction 'to harness the power of heritage to accelerate the achievement of the SDGs' is to mobilise the 'Culture-Nature' approach and landscape-based solutions to achieve the well-being of the Planet, which is related to SDGs 6, 7, 11, 13, 14 and 15 (Labadi et al. 2021). This explicitly recognises the nature-culture nexus on the landscape scale, a feature of biocultural landscapes, as a means of sustainable development and achieving SDGs.

Many discussions are related to the role of heritage conservation in sustainable development. However, they usually only take place at the theoretical level and, in many cases, only focus on the economic component while ignoring the social and environmental dimensions (Courtney et al. 2006; Nocca 2017). Therefore, future directions for heritage conservation that include the biocultural landscape must also include social and environmental development agendas. In the context of the SDGs, heritage conservation must support the achievement of the goals of good health and well-being, gender equality, clean water and sanitation, climate action, life underwater and life on land. Thus, heritage conservation approaches must be broad in scope to include both the within and/or wider surroundings of the heritage site. Therefore, there is an urgent call to refocus on the management and conservation of natural and cultural heritage sites, accounting for the wider context of the site and options for its sustainable use (Engels 2015). For this reason, the concepts of landscape and heritage site are closely related: heritage can exist within and influence the landscape, and the landscape can be managed as a heritage site (Vileniske 2008). In the future, planning for heritage site management and conservation must be done with a spatial approach without abandoning the historic landscape itself (Caballero 2016; Bonenberg 2020).

Generally, the conservation of natural and cultural heritage is already contained within the sustainable development framework, which has three interrelated pillars of sustainability: environment, economics and society. In this regard, the keywords for heritage conservation have shifted from environment and protection to emphasis on the sustainable development agenda (Rodwell 2012). This move is crucial for conserving a biocultural landscape, as it provides a pathway for enhancing sustainable development, especially in the highly changing environments of developing economies like Malaysia and Indonesia.

15.3 Importance of Conserving Biocultural Landscape for Sustainable Development in Malaysia and Indonesia

The long-time mutual interdependence between humans and the environment and peoples' resource sustainability practices are fundamental features of biocultural landscapes. Explicitly and implicitly, the feature provides significant social, economic and environmental benefits, encompassing the three main pillars of sustainable development. In this regard, conserving biocultural landscapes supports the sustainable development agenda, especially for countries like Malaysia and Indonesia, with their rich biodiversity, culture and traditional heritage.

In the ecological context, biocultural landscape conservation promotes biodiversity conservation and enhanced ecosystem services (e.g. Assandria et al. 2018). Biodiversity is an important element of biocultural landscapes. For the local community, biodiversity is essential as it is a source of food as well as income generation. Sustainability approaches using traditional ecological knowledge to use biological resources play an important role in maintaining the availability of resources, both in the present and the future. Thus, biocultural landscape conservation promotes sustainable biological resource practices, which consequently enhance biodiversity conservation. In addition, biocultural landscape conservation also complements governmental conservation efforts (if any) in adjacent wilderness areas, such as natural forests that tend to be adjacent to biocultural landscapes.

Biocultural landscape conservation may prevent or reduce the impact of human disturbance to natural or semi-natural wildlife habitats. Other existing traditional land uses, such as croplands, farmlands, orchards and traditional settlements, are also less threatened by development; this may maintain or possibly improve the heterogeneity of the biocultural landscape. Filippi-Codaccioni et al. (2010) revealed that increased heterogeneity could improve the biodiversity of traditionally managed agricultural landscapes. Therefore, heterogeneity enhances biodiversity and improves the 'naturalness' of the biocultural landscape. This heterogeneity attracts more wildlife to natural or semi-natural habitats within the biocultural landscape for foraging and breeding. This process improves the ecological stability and integrity of biocultural landscapes. It increases resiliency and reduces vulnerability to any changes or disturbances caused by, for example, heavy rains, storms, droughts or other natural disasters, as well as phenomena like El Nino and La Nina, and climate change.

Improving the resiliency, stability and integrity of the biocultural landscape ecosystem can also affect the local community's well-being, including their socioeconomics and health. Accordingly, improving ecosystem services allows local people to more easily obtain resources (food or non-food); this is particularly significant if the local community depends on natural resources for their livelihood. In addition, benefits from improved ecosystem services complement their main source(s) of income, such as crops and livestock husbandry. As a result, the local community gains additional sources of income, which can alleviate or reduce poverty.

In addition, agricultural and crop productivity can also increase with improved ecosystem services. For example, the hydrological cycle can be improved, with river systems eventually having better water quality and more volume. Altogether, hydrological systems are important to the agricultural irrigation system, improving crop fertility and production. As such, these improvements can increase the agricultural income of farmers, which is particularly important in the context of food security and climate change. Other sources of income can also be generated through tourism. Since biocultural landscapes are rich in natural and cultural heritage in addition to traditional values, their conservation may encourage nature- and culture-based tourism, favouring the conservation and preservation of biodiversity and cultural diversity. In addition, nature- and culture-based tourism provide opportunities for the local community to generate extra income via new jobs.

A stable ecosystem, abundant natural resources, adequate food supply and a stable income will result in a physically and mentally healthier society. Mental health also depends on the surrounding environment, and biocultural landscape conservation may produce a more peaceful, clean and aesthetic atmosphere. When the well-being of a community and the environment improves, it can also prevent urban migration, especially amongst the young generation, which is an issue in biocultural landscapes (e.g. Rescia et al. 2010; Kamada 2017). If urban migration prevails, the number of elderly individuals may exceed the number of young individuals to a point where the biocultural landscape population becomes an ageing population. This imbalance between elderly and young populations can eventually affect the sustainability of the biocultural landscape. For example, urban migration can cause many agricultural areas to be abandoned as elderly individuals become physically unable to efficiently work their crops, reducing productivity and affecting the income of the remaining residents.

The rich culture, customs and traditional knowledge and wisdom in biocultural landscapes have long persisted. For example, local communities have usually implemented traditional methods or approaches that encompass sustainability into their agricultural activities and land management. Examples of these methods are the use of organic fertilisation (e.g. manure and plant compost) and pest control (e.g. natural prey–predator species) and terrace systems in agriculture to prevent soil erosion. In addition, culture and customs that usually involve revelries and/or festivities (e.g. wedding ceremonies) and daily practices that use resources or social relations emphasise the importance of harmonious cooperation, mutual understanding and respect. In this regard, harmony exists between humans and the environment and between humans, which reflects the property of sustainability. Therefore, biocultural landscape conservation can foster harmony between humans as well as between humans and the environment.

At present, many traditional cultures or practices are in decline due to the influence of modernity permeating into biocultural landscapes. Modernity mostly affects the younger generation, in contrast to the elderly generation who mostly still adhere to traditional customs or practices. Nonetheless, through biocultural landscape conservation, strategies can be formulated to ensure the continuity of traditional culture and practices; however, these strategies must be consistent with the current situation and suit the spirit of modernity, which is in the heart of young people. This approach is necessary so the younger generation understands that their presence in the biocultural landscape community is valued and recognised, which may avoid or reduce urban migration. It would inspire younger generations to appreciate and value that traditional heritage remains relevant in today's world.

15.4 Conclusions

In Malaysia and Indonesia, biocultural landscapes are places or localities with heritage value. The closely interrelated elements of nature and cultural practices are at the core of the characteristics of a biocultural landscape. Nevertheless, identifying places or areas as biocultural landscapes is still in its infancy in both countries. As countries rich in natural resources and diverse cultural backgrounds, many places, especially in rural areas, can be characterised by their biocultural landscapes. Current efforts to highlight the privileges and role of biocultural landscapes in achieving sustainable development are intensifying. This is mostly done through research activities in identified areas characterised by a biocultural landscape.

Research aims to curb the growing threat of development in biocultural landscapes of the countries. In this case, economic development and population growth are the driving factors that threaten the conservation of biocultural landscapes. Thus, research results provide important basic data in support of the need for biocultural landscape conservation. More importantly, the conservation of biocultural landscapes is a platform to increase the seriousness of conserving the resources and values of natural and cultural heritage through a more holistic approach. To date, a holistic approach to conserving the heritage of natural resources and culture has been adapted and emphasised by relevant international bodies such as the UNESCO World Heritage Convention, ICCROM, ICOMOS and IUCN. The conservation of these natural and cultural heritage resources has become an important effort in supporting the global sustainable development agenda through these

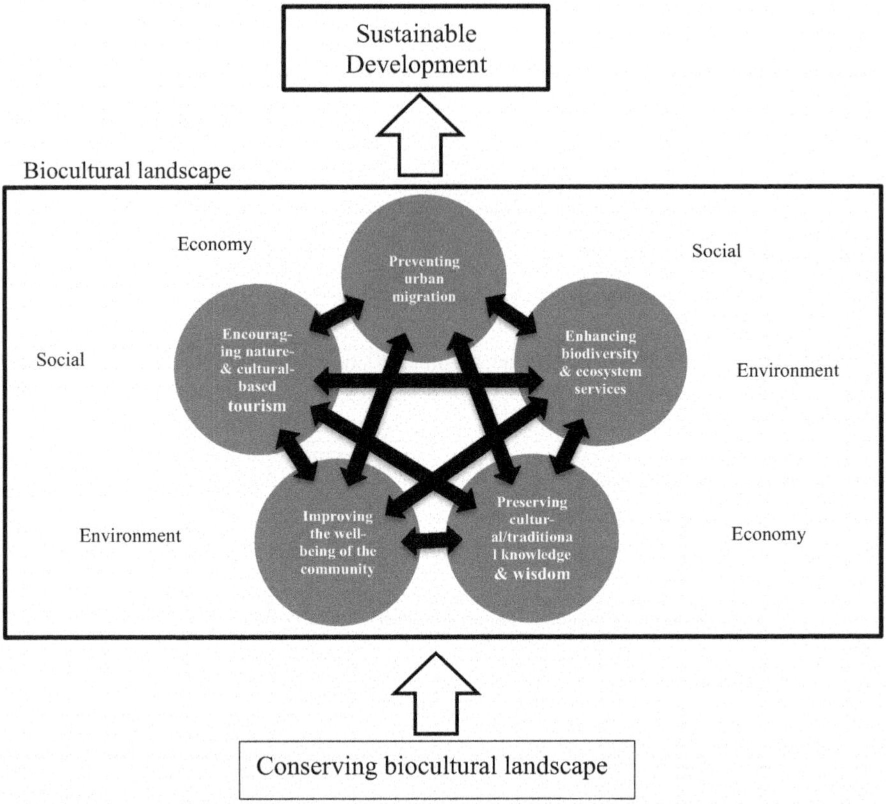

Fig. 15.1 Relationships demonstrating the importance of conserving biocultural landscapes as necessary in support of sustainable development in Malaysia and Indonesia

international bodies. From this point forward, there is a need to conserve the biocultural landscape for sustainable development and achieve the SDGs by 2030.

In this context, Malaysia and Indonesia can promote biocultural landscape conservation in support of the sustainable development agenda of their specific respective countries and the southeast Asia region in general. In this regard, we have outlined and identified the main points that highlight the importance of conserving biocultural landscapes. Each point is interrelated, namely strengthening biodiversity conservation and ecosystem services, preserving cultural or traditional knowledge and wisdom, improving the community's well-being, preventing urban migration and encouraging nature- and culture-based tourism (Fig. 15.1). The interrelated points lead to sustainable development, in which the benefits fall within the three aspects of sustainability: environmental, economic and social. This effort requires the support and cooperation of all stakeholders, especially at the policy-making level, to achieve sustainable development.

References

Assandria G, Boglianib G, Pedrinia P, Brambillaa M (2018) Beautiful agricultural landscapes promote cultural ecosystem services and biodiversity conservation. Agric Ecosyst Environ 256:200–210

Bonenberg W (2020) The role of cultural heritage in sustainable development. Values and valuation as key factors in spatial planning of rural areas. In: Charytonowicz J, Falcão C (eds) Advances in human factors in architecture, sustainable urban planning and infrastructure, AHFE 2019. Advances in intelligent systems and computing, vol 966. Springer, Cham. https://doi.org/10.1007/978-3-030-20151-7_12

Caballero GVA (2016) The role of natural resources in the historic urban landscape approach. J Cult Herit Mgmt Sustain Dev 6(1):2–13. https://doi.org/10.1108/JCHMSD-11-2014-0037

Chapagain NK (2019) Towards bio-cultural landscape: a case of Kutch. In: Chiba M, Jain S, Ghosh S, Mathur VB (eds) Cultural landscapes of Asia: understanding and managing heritage values. Dronah publisher, pp 153–159

Courtney P, Hill G, Roberts D (2006) The role of natural heritage in rural development: an analysis of economic linkages in Scotland. J Rural Stud 22(4):469–484. https://doi.org/10.1016/j.jrurstud.2006.02.003

Elliot JA (2006) An introduction to sustainable development. Routledge, London, 283p

Engels B (2015) Natural heritage and sustainable development – a realistic option or wishful thinking? In: Albert M (ed) Perceptions of sustainability in heritage studies. De Gruyter, Berlin, pp 49–58

Filippi-Codaccioni O, Devictor V, Bas Y, Clobert J, Julliard R (2010) Specialist response to proportion of arable land and pesticide input in agricultural landscapes. Biol Conserv 143:883–890

Kamada M (2017) Satoyama landscape of Japan—past, present, and future. In: Hong SK, Nakagoshi N (eds) Landscape ecology for sustainable society. Springer, Japan, pp 87–109

Labadi S (2014) UNESCO world heritage convention (1972). In: Smith C (ed) Encyclopedia of global archaeology. Springer, New York. https://doi.org/10.1007/978-1-4419-0465-2_1039

Labadi S, Giliberto F, Rosetti I, Shetabi L, Yildirim E (2021) Heritage and the sustainable development goals: policy guidance for heritage and development actors. ICOMOS, Paris

Nocca F (2017) The role of cultural heritage in sustainable development: multidimensional indicators as decision-making tool. Sustainability 9(10):1882. https://doi.org/10.3390/su9101882

Rescia AJ, Willaarts BA, Schmitz MF, Aguilera PA (2010) Changes in land uses and management in two nature reserves in Spain: evaluating the social–ecological resilience of cultural landscapes. Landsc Urban Plan 98:26–35

Rodwell D (2012) The UNESCO world heritage convention, 1972–2012: reflections and directions. Hist Environ 3:64–85

Slatyer RO (1983) The origin and evolution of the world heritage convention. Ambio 12:138–140

UNESCO (1972) Convention concerning the protection of the world cultural and natural heritage. UNESCO, Paris

UNESCO (1994) Operational guidelines for the implementation of the world heritage convention. UNESCO, Paris

UNESCO (2002) Convention concerning the protection of the world cultural and natural heritage. UNESCO, Paris. World Heritage Committee 26th session

UNESCO (2007) Convention concerning the protection of the world cultural and natural heritage. UNESCO, Paris. World Heritage Committee 31th session

UNESCO (2011a) Convention concerning the protection of the world cultural and natural heritage. UNESCO, Paris. World Heritage Committee 35th session

UNESCO (2011b) Convention concerning the protection of the world cultural and natural heritage. UNESCO, Paris. 18th Session of the General Assembly of the States Parties

UNESCO (2015) Convention concerning the protection of the world cultural and natural heritage. UNESCO, Paris. 20th Session of the General Assembly of the States Parties

Vileniske IG (2008) Influence of built heritage on sustainable development of landscape. Landsc Res 33(4):425–437. https://doi.org/10.1080/01426390801946491

The manufacturer's authorised representative in the EU is Springer Nature Customer Service Centre GmbH, Europaplatz 3, 69115 Heidelberg, Germany. If you have any concerns regarding our products, please contact ProductSafety@springernature.com

Printed and bound by CPI Group (UK) Ltd, Croydon, CR0 4YY
30/09/2026
02225233-0002